D1242913

Remarkable Birds

PELECANUS Carbo.

– *Remarkable* –

BIRDS

Mark Avery

225 illustrations, 203 in color

Thames & Hudson

Front cover: American White Pelican,
John James Audubon, *The Birds of America*, 1827–38.
Back cover: Hummingbirds,
Robert John Thornton, *Temple of Flora*, 1807.

Half-title: Eggs of Great Tit (left) and Raven (centre and right).
Frontispiece: Cormorant, Cornelius Nozeman, *Nederlandsche Vogelen*, 1770–1829.
This page: Skylark, Cornelius Nozeman, *Nederlandsche Vogelen*, 1770–1829.

Remarkable Birds © 2016 Thames & Hudson Ltd
Text © 2016 Mark Avery
Illustrations © 2016 The British Library Board,
unless otherwise stated, see pp. 235–6

Mark Avery has asserted his moral right to be
identified as the author of this work.

All Rights Reserved. No part of this publication may
be reproduced or transmitted in any form or by any means,
electronic or mechanical, including photocopy, recording or
any other information storage and retrieval system, without
prior permission in writing from the publisher.

First published in 2016 in hardcover in the
United States of America by Thames & Hudson Inc.,
500 Fifth Avenue, New York, New York 10110

thamesandhudsonusa.com

Library of Congress Catalog Card Number 2015959511

ISBN 978-0-500-51853-3

Printed and bound in China by C & C
Offset Printing Co. Ltd

Contents

CENTRAL ARKANSAS LIBRARY SYSTEM
ADOLPHINE FLETCHER TERRY BRANCH
LITTLE ROCK, ARKANSAS

The Love Life of Birds *86*

Avian Cities *116*

Useful to Us *140*

Threatened & Extinct *164*

Revered & Adored *196*

1

Introduction

2

We share the Earth with more than 10,000 species of birds. They soar over the highest peaks and skim over the ocean waves. Their songs fill the rainforests and yet we also wake to their melodies in the middle of our biggest cities. Almost every human who has ever lived has probably seen or heard a bird almost every day. The presence of birds has been a comfort to travellers, a tonic for the ill and a delight to many as they go about their busy lives.

We like birds: we feel serenaded by their songs, we marvel at their mastery of the air, we approve of their attention to their young, and, in certain cases, we think they taste delicious too. And birds seem to treat us with respect – as we approach them they usually fly away rather than attack us or stand their ground. Some mammals may eat us, some snakes may bite us and some insects may sting us, but most birds look pretty, sound melodious and generally defer to us as we go through life. Birds are easy companions for us and can amuse and soothe us.

We also experience the world in a similar way to many birds, which adds to our feeling of kinship. Birds are creatures of sight and sound, as are we, whereas most of our fellow mammals' sensations are dominated by smell and touch. We admire the bright, sometimes gaudy plumage of many male birds, and realize that so do the females of those species, and we delight in the brilliance of the finest avian songsters, and appreciate that those songs are sending messages to other birds. Birds build nests that are their temporary homes, and raise their young by guarding them from danger, keeping them warm and feeding them. They please us, intrigue us and sometimes astonish us with what they can achieve.

Birds evolved from dinosaurs – in fact we can think of them as feathered dinosaurs. The fossil *Archaeopteryx* (from *archaios*, meaning 'ancient', and *pteryx*, meaning 'feather' or 'wing'), discovered in Germany in 1861, dates from around 150 million years ago. This early bird, which was 50 cm (20 in.) long, was covered in feathers and had broad wings and tail.

3

It was warm-blooded and could certainly glide, and perhaps fly weakly as well; its small breastbone suggests that its flight muscles would not have been very well developed. Other proto-bird fossils have now also been identified, which push back the first known fossil feather to 160 million years.

In those 160 million years, birds have radiated into the 10,000-plus species we see, and hear and love today. The largest surviving bird species is the Ostrich, which inhabits the plains of Africa and Arabia. It is taller than a human and, at 100 kg (220 lb), is fifty thousand times heavier than the smallest bird, the Bee Hummingbird, which weighs in at a little below 2 g (0.07 oz).

The polar wastes support very few birds – although birds have been seen at both the North and South Poles – while the tropical regions are richest in bird species. South America hosts 36 per cent of all land-bird species, Africa 21 per cent and Southeast Asia another 18 per cent. These figures are reflected in the numbers of bird species found in individual countries. Colombia, far from the largest country on Earth (in fact only twenty-sixth largest), has more bird species than any other – more than 1,800 – and its much larger neighbours, Peru and Brazil, are second and third on the list.

4

Every one of these bird species on our planet is remarkable in its own way, and here we have space to cover only around 67 of them. These are grouped into eight categories so that we can explore certain themes and connections in our appreciation of, and relationship with, birds, and linger over aspects of their lives in greater detail in order to celebrate them more fully. Of course, many could have fitted into more than one of our categories.

5

Bird song is a joy to our ears, although the early morning crow of the cockerel may be greeted with only grudging delight. In 'Songbirds' we celebrate the beauty and variety of bird song, but also investigate its functional significance. We may pause as we go about our daily chores to listen to the sound of a singing bird, but song is energetically costly (as any chorister will confirm), and the male bird who pours his heart out is actually defending his territory and attracting a mate, both at the same time. For birds, song is deadly serious – song is about sex and violence, not about joy and rapture. And yet, for us, their songs of challenge and lust are often of sublime beauty. Some birds, rather than being melodious, excel in being great mimics – of other birds, but also increasingly of human-made sounds, even voices.

'Birds of Prey' have always been admired by mankind. They are hunters, as were our human ancestors, and they have evolved specialist skills and often incredible eyesight to be able to hunt other birds, mammals, fish and even molluscs. These are birds of majesty, whose command of the air excels that of most other birds.

The ability to fly not only allows birds to move around quickly and directly, and to seek safe places to shelter and nest, but also means they can be great travellers. Flight enables birds to migrate immense distances and overcome barriers such as seas, deserts and mountain ranges that would be impractical on foot. 'Feathered Travellers' describes examples of the most astounding journeys made by birds, and some of this information is only now being revealed through the use of modern technology.

6

7

Most bird species form pair bonds, with a male and a female raising and tending their young in a partnership that lasts for at least a season, but often for life – although, as with our own species, these relationships are not always as fully respected as might appear to the outside world. Occasional infidelity is commonplace among bird species, but in some the males then play no part in looking after the young, and all their efforts go into attracting numerous mates. Such species have the most spectacular displays and the most extravagant plumages of all birds. 'The Love Life of Birds' describes the variety of conjugal arrangements among birds, which can rival those of any TV soap opera.

Although the usual pattern is for a pair of birds to establish a territory, which they defend against others of their own species and in which they rear their young, around one in ten bird species live in colonies. This occurs where either nest-sites or food resources, or both, are limited, and birds are crowded together in great avian metropolises. 'Avian Cities' explores colonial species and the advantages and disadvantages of living in close proximity with so many neighbours. Large colonies of birds provide us with some of the greatest wildlife spectacles on the planet.

8

'Useful to Us' examines the ways in which we have found birds of value to us practically and economically. The most obvious example is the fact that we eat billions of them each year, mostly from a very few domesticated species that are now farmed – yet they came from wild species that are still living naturally. There are some species that we hunt for pleasure, as well as for food, and others that we even use to help us find food.

❾

Many of the world's birds are threatened with extinction – indeed some have already succumbed. 'Threatened & Extinct' describes some of the birds you can no longer find on Earth and others that seem on the brink of following them. But there are also tales of species bouncing back from the edge of oblivion, and these illustrate both our own ability to avert extinction and the resilience of bird species to recover from very low numbers.

From the earliest times, birds have been seen by us as harbingers, sometimes of birth and sometimes of death. Their appearance in the sky, or their calls on the air, have been associated with both celebration and warnings. All over the world, birds have been entwined with human culture in ways that varied with the personalities of the birds and the human cultures which share this planet. 'Revered & Adored' considers the various symbolic and mystical aspects of our relationship with birds.

Our fascination with birds and this relationship in all its aspects also find expression in the great art and literature that birds have inspired, throughout the centuries and around the world, fine examples of which can be found in these pages.

❶ The Wild Turkey, from which the edible turkey was domesticated, is a surprisingly striking bird. ❷ The sound of the Cuckoo in northern climes heralds the return of spring. ❸ Many myths surround the enigmatic Wandering Albatross. ❹ Hummingbirds are among the smallest birds on the planet, yet some migrate immense distances. ❺ The Golden Eagle is the king of birds, and has been used to hunt wolves. ❻ The Bar-tailed Godwit makes a non-stop eight-day flight on its autumn migration from Alaska to New Zealand. ❼ The exotic-looking Hoopoe has been considered sacred by several different cultures. ❽ The chicken, related to the wild Red Junglefowl, is today the most numerous species of bird on Earth. ❾ In Phalaropes, unusually, it is the female that is the more brightly coloured bird.

Songbirds

Nightingale • Skylark • Wood Thrush • Goldcrest • Eastern Koel
Great Tit • African Grey Parrot • European Starling
Lyrebirds • Northern Mockingbird

ABOVE Great Tit.

We give the name 'song' to many birds' vocalizations. When we sing, it can be an emotional experience for both performer and audience. Bird song touches our hearts in a similar way, and we ascribe emotions of love, loss and longing to the singer. There is no musical feast that can rival the songs of the world's best songbirds. And they sing to us for free.

In early human history, long before machine-made noise, the songs of birds were the main sounds, and pleasing ones, to punctuate the background soundscape of wind and waves. Our ancestors would often have wakened to a great chorus of bird songs – a blend of many voices of numerous species in which the individual contributions were hard to distinguish. But some birds are renowned for their exceptional vocal abilities, and poets have long celebrated the songs of the Nightingale and Skylark for their beauty and virtuosity – and rightly so.

Of course, birds don't sing for our benefit: they sing to attract mates and to defend their territories. Bird song is mostly, therefore, heard during the breeding season, and in higher latitudes this means in spring. Birds such as the Eastern Koel in Australia and the Wood Thrush in North America are regarded as harbingers of spring because their songs sound joyous and hopeful, and indicate that longer days and warmer weather are ahead.

But there is more to it than that, and we now more fully understand the role that song plays in the lives of birds. Although it might seem to be one of the simplest songs of any bird, the Great Tit song is highly complex once examined in detail. Some birds mimic the songs of others, indicating that songs are learned as well as being hard-wired into the birds' brains. The Lyrebird, European Starling and Northern Mockingbird all mimic other species' calls and sometimes incorporate mechanical noises of human origin into their songs. The African Grey Parrot is also an accomplished mimic of human vocalizations.

And we should not fool ourselves that, however hard we listen, we hear all of a bird's song: some songs, such as that of the Goldcrest, are partly pitched too high for the human ear to hear them.

NIGHTINGALE

Luscinia megarhynchos

There are few birds that are such virtuoso songsters as the Nightingale, a bird that is commonly found across much of Europe, North Africa, the Middle East and Southwest Asia. The appreciation of the Nightingale's song is enhanced by the fact that it is loud, and although it can be heard during the day, it is most often at night that it can be enjoyed, when other bird song has ceased.

The Nightingale itself is a rather drab brown bird and resembles the slightly smaller Robin (*Erithacus rubecula*); it tends to skulk in thick vegetation, feeding on insects and fruit, so it is rarely seen.

ABOVE
While undistinguished in plumage, the Nightingale is one of the most accomplished avian songsters. On spring evenings, as other bird song comes to an end, male Nightingales fill the woods with loud musical phrases to defend their territories and attract mates.

OPPOSITE
Nightingales nest on the ground, usually among leaf litter. The clutch of four or five eggs is incubated by the female for 14 days, and the young are fed in the nest by both parents for a further 11–13 days.

Nightingales are common in the Mediterranean areas, and it is therefore no wonder that Sophocles, Ovid, Virgil, Aristophanes, Callimachus and Homer knew this bird and admired its song. Ovid wrote of the Nightingale as Philomela, a wronged woman:

My mournful voice the pitying rocks shall move,
And my complainings echo thro' the grove.
Hear me, o Heav'n!

And yet it is male Nightingales that sing, in snatches of a few seconds punctuated by pauses of the same length. Within each snatch of song there may be multiple repeats of tuneful whistles or churring notes. The frequent musical trills have over 100 musical elements per second and are tiring for the birds. Not surprisingly perhaps, older males sing faster trills with a broader sound frequency width than younger ones, because practice makes perfect. It is likely that female Nightingales can judge the age and physical health of a singing male by characteristics of its song. Males lose weight as they sing each night – it

1
2

RIGHT
The song of the Nightingale has inspired poets over the centuries. Some have interpreted it as joyful, others as an outpouring of regret or loss. In fact, the song warns other males to steer clear and is used to attract females.

OPPOSITE
When they leave the nest, young Nightingales are a little more streaked than their parents, but they are indistinguishable from older birds on their return from their wintering grounds in Africa. The songs of young males are slightly less polished than those of experienced males.

is an energetically demanding but necessary process for defending territories and attracting mates. Pauses in the song give the impression that the bird is a performer, wondering what to extemporize next. Indeed, male Nightingales each have around 200 different snatches of song to choose from, with older males again having more varied songs than young birds.

But rather than over-analyse the wonderful song of the Nightingale, we should celebrate it. When Nightingales return in spring to Europe and Asia from their African wintering grounds (in a broad band from Senegal to Tanzania), they sing for a period of a little over a month. Listening in woodland to the outpouring of song from an unseen but nearby Nightingale in the dusk or dark, with little other noise apart from a distant hooting owl, is an experience that can be immensely moving.

It is the impression of skill and joy that makes a Nightingale's performance so captivating. As John Keats wrote:

... thou, light-winged Dryad of the trees,
In some melodious plot
Of beechen green, and shadows numberless,
Singest of summer in full-throated ease.

SKYLARK

Alauda arvensis

The Skylark sings while hovering high in the sky, way above our heads. It is the male that sings and one function of his song is to proclaim his territory and keep out rivals. Often, while its song is clearly heard, the bird itself is difficult to discern – just a small dot in the sky, held there by its rapidly beating wings as the liquid notes spill out over the fields below.

The English poet Percy Bysshe Shelley captured this perfectly in his ode 'To a Skylark':

Hail to thee, blithe spirit!
Bird thou never wert—
That from heaven or near it
Pourest thy full heart
In profuse strains of unpremeditated art.

And later, in his poem 'The Lark Ascending', George Meredith added his paean of praise to this accomplished songster:

He rises and begins to round,
He drops the silver chain of sound,
Of many links without a break,
In chirrup, whistle, slur and shake.

The collective name for a group of Skylarks is an 'exaltation', although, being a territorial species, one rarely sees a group of Skylarks together except in the winter when they flock, but at that time they only make their warbling calls rather than the fully exalted song. On a fine spring day, it is still possible to listen to several Skylarks scattered across the skies over the landscape, seemingly filling the air with their flowing notes. However, Shelley and Meredith,

ABOVE
Skylarks lay clutches of three to six eggs, making their first attempts to nest in late April. Several nesting attempts are made depending on weather conditions and food availability.

OPPOSITE
Made of grass stems, the nest is built on the ground and is home to young Skylarks until they are around 11 days old. Nests are hidden in tall vegetation, often grasses or cereal crops, but dense greenery is shunned as it makes access to the nest too difficult for the adults.

Oscar Dressler dis e lit

ABOVE
Skylarks feed their young on insects, but through the year have a varied diet that includes seeds, vegetation and a range of invertebrates. Although they form flocks in winter, in spring individual males can be heard pouring out their songs as they hover high above the fields in which they feed and nest.

and even our fathers and mothers, would have heard many more Skylarks than the present-day walker through the European countryside, as the Skylark has declined considerably in numbers due to the impacts of modern farming.

Larks are birds of open country – meadows, plains, heaths and now arable farmland and intensively managed grassland – both nesting and feeding on the ground, with their diet consisting mainly of seeds and insects. Unfortunately, the intensive management of grass, with mechanized and earlier cutting, and the thick crops of fast-growing cereal which are sprayed with herbicides, are far from perfect nesting and feeding sites for Skylarks. Skylarks can raise as many as three broods of young in a season, laying three to six eggs each time, but rarely manage more than one in modern arable crops. Thus, the population trend for the Skylark has plummeted, just as a Skylark falls to earth after a bout of singing high above our heads.

Many other lark species (there are nearly 100 species altogether) are facing similar difficulties, and it has seemed as though these characteristic birds of European and Asian open landscapes were destined to decline in numbers with the intensification of modern agriculture. Can we have exalting larks and cheap food produced by the same land? Fortunately, there are signs of hope as scientists have developed a simple means of enabling Skylarks to cope with modern farming practices. Leaving small unsown patches, not much bigger than a dining table, scattered through fields of wheat or barley provides landing strips, feeding plots and sometimes nesting sites for Skylarks. Fields where these small patches have been left by farmers retain their Skylarks, and it has been shown that their nesting success is high.

There is arguably nothing as redolent of spring and early summer as to lie back in long grass and listen to the sound of several Skylarks singing way up in the sky (or even just to imagine it). Their songs fill the air and the senses like no other bird. At times like that, the phrase 'exaltation' seems entirely appropriate, and you will head homewards with your own spirits lifted by the beautiful song of a small brown bird.

WOOD THRUSH

Hylocichla mustelina

ABOVE
The flute-like notes of a Wood Thrush herald the return of spring to the deciduous woods and forests of eastern North America. Wood Thrushes winter in the rainforests of Central America, with the males returning north in April, ahead of the females.

T. S. Eliot celebrated the Wood Thrush, whose beautiful fluty song fills the deciduous woods of eastern North America each spring, in the opening and closing lines of his poem 'Marina' (published in 1930), describing it 'singing through the fog'. And earlier, in June 1853, Henry David Thoreau wrote of the Wood Thrush '... singing his evening lay. This is the only bird whose note affects me like music, affects the flow and tenor of my thought, my fancy and my imagination. It lifts and exhilarates me. It is inspiring.... It changes all hours to an eternal morning.'

Although they are not very closely related species, the Wood Thrush does resemble both in plumage – brown above and with a gorgeously speckled breast – and in the beauty of its song, the Song Thrush of Eurasia (*Turdus philomelos*), which has also been lauded by poets. The Wood Thrush's song consists of short musical phrases, usually of three notes: 'ee – oh – lay', separated by silence of a few seconds. The notes are clear and are often described as flute-like in quality.

Birds of woodlands as their name suggests, they often feed in the leaf litter of the forest floor for small invertebrates, mainly insects but also earthworms, millipedes and snails. Snails are an important part of the diet as they provide the calcium that the females need to make their eggshells. As much as 6 per cent of the weight of the egg is calcium, and that takes some finding in order to lay a clutch of two to four pale blue eggs. Numbers of Wood Thrushes have declined by 40 per cent since Eliot mentioned them, and acid rain leading to a reduction in the abundance of forest-living snails might be one of the causes behind that decline.

Wood Thrushes are migrants, overwintering in Central America, in countries such as Honduras and Nicaragua in areas where forest clearance is at a

1
2

high level – and so habitat loss in the wintering grounds may be another factor driving the decline in the breeding population.

The males return to their woods in the north in late April, a little before the females, and start singing to defend territories and attract mates. The male Wood Thrush is an attentive father and does more than half of the provisioning for the young nestlings, although genetic analysis has shown that up to 40 per cent of the young in any nest may be fathered by males who are not paired with the resident female. Two broods may be raised each year.

OPPOSITE
The Wood Thrush would have been a familiar species to celebrated American ornithologist John James Audubon, both at his early home in Pennsylvania and later as he produced his *Birds of America* (1827–38) in Kentucky. He described it as 'my greatest favorite of the feathered tribes of our woods'.

BELOW
Wood Thrushes share their deciduous woodland nesting habitat with a wide range of other migrant songbirds, including the smaller Ovenbird (*Seiurus aurocapilla*) pictured here.

A further difficulty in the Wood Thrushes' lives is nest parasitism by Brown-headed Cowbirds (p. 108). Cowbirds lay their eggs in the nests of other species and they or their chicks eject the existing eggs or nestlings. Generally preferring open country, Cowbirds formerly found host species accordingly, but forest fragmentation in North America has meant that many forest-dwelling species are now more vulnerable. Some bird species, perhaps from experience, are good at spotting the Cowbird eggs and eject them from their nests, but Wood Thrushes, at least to date, have not evolved such perspicacity and discrimination; they accept the eggs and are now a favoured host species.

Arthur Cleveland Bent wrote in 1949 that 'The nature lover who has missed hearing the musical bell-like notes of the wood thrush, in the quiet woods of early morning or in the twilight, has missed a rare treat. The woods seem to have been transformed into a cathedral where peace and serenity abide. One's spirit seems truly to have been lifted by this experience.' Each year, Americans eagerly await the return of the song of the Wood Thrush for its beauty and also as a harbinger of spring. The Wood Thrush's songs still resonate through the stillness of the early morning, but because of their population decline, their voices are far less numerous than formerly.

GOLDCREST

Regulus regulus

ABOVE
This tiny bird is often seen hovering at the tops of conifer trees, gleaning insects from the foliage. The male (right) has an orange centre to the golden crest, which the female (left) lacks. When the bird is agitated, the feathers of the crest are raised to display its bright colour fully.

The Goldcrest is the smallest European bird and its song is so high-pitched that it can be inaudible to some people. Its name comes from the yellow crest which it raises in alarm or display, and which in the male has an orange centre. It weighs just 5 g (0.2 oz) and is very vulnerable to cold winters, when populations can crash.

The song is a series of double notes repeated several times, with the whole song repeated several times a minute. Goldcrests sing frequently and with enthusiasm, often high in the tops of dense conifer trees, where they feed on insects. The songs of some birds are used by birdwatchers and scientists to monitor particular populations, as many species, like the tiny Goldcrest, are more easily heard than seen.

In legend, too, Goldcrests have a reputation for hiding. Aristotle wrote that the birds were competing to be king and it was decided that the one that flew highest would receive the title. The eagle soared higher than all the others, until a small bird, by some said to be a Wren but by others a Goldcrest, flew from its hiding place in the eagle's feathers and hovered above it.

The Goldcrest song is so highly pitched that it is one of the first songs that older birdwatchers may begin to fail to hear. Some bird monitoring schemes suggest that Goldcrests are declining, but an alternative explanation is that the community of bird surveyors may have ageing ears.

EASTERN KOEL

Eudynamys orientalis

ABOVE
The female Eastern Koel (top) has brown upperparts spotted with cream, and cream underparts streaked with brown; the male (below) is glossy black with a red eye. The arrival of this bird in Australia from its northern wintering grounds, announced by the distinctive song of the males, signals the imminent wet season.

For people living in eastern Australia, the distinctive sound that heralds the coming of the austral spring and the rainy season is an oft-repeated, loud and ascending 'coo-ee' or 'ko-el' song. In this way, the Eastern (or Common or Pacific) Koel announces its arrival in the coastal forests in September/October, just as the Cuckoo, to which it is related, does in the northern hemisphere spring. The song is delivered over and over again through the daylight hours, with peaks at dawn and dusk, but also sometimes after dark.

The birds are returning to Australia to breed, and the males are defending their territories. They arrive a week or so before the females, which then add their strident 'keek keek keek keek' calls to the sounds of the forest, but also the edges of towns and cities as this species is increasingly found, and heard, in suburban gardens. The males are easy to hear but not so easy to see, despite sitting high in the canopy of a tall tree, where this species feeds on fruit. Males are glossy black all over, with a bright red eye. The females, who lay their eggs in the nests of other species of bird, mainly the Red Wattlebird (*Anthochaera carunculata*), are spangled brown above and barred toffee-coloured below.

The Eastern Koel is sometimes known as the Rainbird or Stormbird because, as well as arriving before the seasonal rains, it is thought to sing more frequently just before the onset of rain showers. It then departs for its northern wintering grounds in Papua New Guinea and Indonesia in March and April, when the rainy season comes to an end.

GREAT TIT

Parus major

A familiar woodland bird and frequent visitor to gardens in Europe, the Great Tit has a wide range, occurring across Eurasia and parts of North Africa. It has an apparently simple and repetitive song, which most books render as 'Tea-cher, tea-cher, tea-cher', and that is pretty accurate. It's a common sound of spring and an easily recognized part of the avian chorus.

And yet, the simple song of the Great Tit, when studied by scientists, has shown us how complex and varied bird song actually is, and also how birds use and perceive their own songs. The Great Tit's song has a dual purpose – to attract mates to, and drive away other males from, a male's own territory (a portion of woodland in which he and his mate will nest, feed and find the food for their nestlings).

Males certainly sing to attract a mate. In experiments in which female Great Tits were captured and kept out of sight of their mates for a short period (30 minutes), the male Great Tits spent much more time singing than neighbouring birds whose mates were still at large. When the female was released, her male's song rate dropped again to normal. Females will disappear naturally in the breeding season, perhaps when predated by Sparrowhawks or weasels. The removal of a female clearly stimulates the bereft male to sing much more, but whether this is to attract a new mate or to regain contact with his old one, or perhaps both at once (whichever happens first), is not known.

Male Great Tits are also listening to the songs of neighbouring males. Studies have demonstrated that each male has several – between one and nine – slight variants on his song, and also that territorial males recognize the songs of established neighbours. If the neighbouring male makes an excursion from his own territory and sings in another's, the invaded male responds

ABOVE
The acrobatic Great Tit is a common visitor to garden bird feeders, particularly in winter. The width of the black central stripe down the chest and the belly is an indication of dominance in the males: wider means tougher. This bird, however, is a female as the black stripe does not even come close to spanning the whole gap between the legs.

OPPOSITE
Great Tits are hole-nesters, nesting in artificial nest-boxes where provided or in natural holes in trees. They lay around eight eggs, and the hatched young are fed by both parents, largely on moth caterpillars, until they leave the nest at around 20 days old. The parents continue to feed their young as families move through the woods in noisy parties.

Taf. 94

RIGHT
In winter, Great Tits do not defend territories but roam the woods in parties, often mixed-species flocks with other tits, Treecreepers (*Certhia familiaris*) and finches.

BELOW
A female Great Tit lays her clutch of eggs, one a day, over a period of around eight days. Together they weigh about two thirds of her body weight.

OPPOSITE
The Great Tit is bigger than most similar species. Here it is pictured with the Crested Tit (*Lophophanes cristatus*, above), Coal Tit (*Periparus ater*, centre) and Marsh Tit (*Poecile palustris*, below). All live largely in woodlands, feed their young on insects and eat mainly seeds in winter.

PARUS MAJOR

immediately by confronting the intruder with displays and loud and repeated songs. The early season is a time of setting up territory and testing the defences and mettle of your neighbours, after which the invisible edges of the territories form largely respected barriers and each male keeps to his own patch.

Monitoring the songs of neighbouring males may also be a way that male Great Tits can spot opportunities to enlarge their territory. At the same time, they can reassure themselves that no one is encroaching on their own. If an unfamiliar male, with unfamiliar songs, ventures to encroach, he is met with immediate challenge and aggression until driven out.

Great Tits sing more frequently in the early morning than at other times of day, as do many other birds which nest at high latitudes – the dawn chorus is a spectacular feature of spring and early summer mornings. Why might it be that there is this great burst of song at the beginning of the day? Poets may regard it as an outpouring of pure joy, but scientists have more prosaic suggestions. After the long night, male birds may want to check whether their neighbours have made it through the night, and to proclaim the fact that they have. And perhaps the still early morning air is better for the transmission of sound waves than when the air has warmed and is in greater motion? Maybe birds, being warm-blooded, are getting their singing done when they don't have to compete with sounds of insects and amphibians? Or, and there is evidence for this for Great Tits, perhaps feeding rates are lower in the cool early morning for birds that spot insects by their movement – so the dawn is simply a good time to sing because it is a poor time to forage.

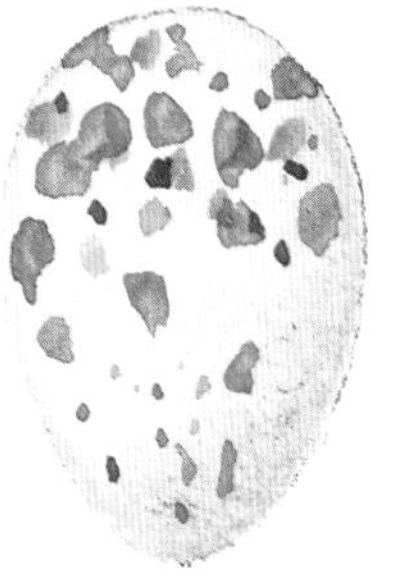

AFRICAN GREY PARROT

Psittacus erithacus

Like several other species described here, the African Grey Parrot is a brilliant mimic, but it is significantly different from the others in one regard: it is mostly silent in the wild. This is not a species with a tremendous vocal repertoire or one that adds in the songs of other birds or other creatures to its song-list. It is instead a rather quiet species that only reveals its ability to mimic when in captivity. Because of this appealing – to humans – characteristic, African Grey Parrots have long been trapped to be traded all over the world and sold as pets, posing a threat to the wild population.

In their native forests of west and central Africa, Grey Parrots are monogamous birds which nest in tree holes, and feed on nuts, fruit and other vegetable matter. As the name indicates, these parrots are almost completely grey, though with a flash of bright red in the tail.

Alex, an African Grey Parrot, was studied in captivity by animal psychologist Dr Irene Pepperburg, who bought him at the age of about one year and named him after her Avian Learning EXperiment. Parrots have often been described by their 'owners' as having great intelligence, but Pepperburg set out to analyse Alex's cognitive and verbal abilities scientifically. At the time of Alex's death in 2007, aged 31, Pepperburg assessed his intelligence as being that of a human five-year-old.

Alex seemed able not just to mimic but actually to talk, with an accurate vocabulary of more than 100 English words for objects, actions and colours. He could ask for food, 'Wanna banana', and would stare silently if offered a nut, and ask again for a banana. He would say 'I'm sorry' and once asked 'What colour am I?' Alex's last words to Pepperburg, but then they were his last words to her every night, were 'You be good, see you tomorrow. I love you.'

OPPOSITE
King Henry VIII of England owned an African Grey Parrot, which would call to boatmen across the River Thames from Hampton Court; they then had to be paid for their trouble. Until recently, when the trade in wild-caught birds was banned in many countries, more than 36,000 African Grey Parrots were traded in the world each year – an indication of their popularity as pets.

Pappagallo cenerino di Guinea, con la coda rossa. — Psittacus Guinensis cinereus, cauda rubra.

All' Ill.mo Sig.re Dott.re Natale Saliceti Pubb.co Profess.re di Anatomia nella Sapienza di Roma, Medico di Collegio, del Palazzo Apostolico, e dell' Arcispedale di S. Spirito ec.

1

EUROPEAN STARLING

Sturnus vulgaris

To see and hear a European Starling sitting on a chimney pot or telegraph wire, with its throat feathers puffed out, its wings making movements like a swimmer doing front crawl, and a series of whirring clicks interspersed with melodious whistles pouring from its open mouth, is a familiar experience for many town and country dwellers. The starling is a common bird in urban areas, woods and the farmed landscape of its native Europe and western Asia, sometimes gathering in huge flocks. It nests in our towns and cities and, despite recent worrying declines in numbers, it is one bird that many people know and recognize. It appears dark at first but in fact has glossy, brightly speckled plumage that is almost iridescent.

Some people love the European Starling; others are not so keen. These birds are pushy, gregarious, noisy and slightly aggressive to each other and to other birds. While some people admire their confidence, others wish they would behave with a little more decorum. They are, perhaps, the avian equivalent of a boisterous group of youngsters out for a night on the town – a little too noisy, a little inconsiderate of others, but basically just getting on with their lives and having a good time. Europeans who travelled to North America, South Africa, Australia and New Zealand took European Starlings with them to remind them of home, and large populations are now established in many parts of the world far from the bird's natural range.

In summer, European Starlings can be seen flying at rooftop height with a prey item, mostly insects, in their pointed, dagger-like beaks, heading back to their nests in a tree hole, a nest-box or perhaps in a gap under the eaves of a house. In winter, European Starlings are frequent and noisy visitors to gardens, where they often gather in crowds in search of food. Although they may squabble and utter aggressive calls, the flock stays together as they feed on our lawns or bird tables. They are fairly omnivorous, and in towns can be seen foraging among litter and discarded takeaways.

OPPOSITE
The European Starling is a striking bird, with its dark plumage marked with flecks and spots of white. Even when stationary, it stands out; it also strides across our lawns and pavements with confidence, and flies through our towns with purpose, its vocal performances demanding our attention.

ABOVE
European Starlings nest in cavities, primarily in trees or houses, and in the crevices under roofs. The clutch consists of four or five pale blue eggs, and the chicks are often fed on the larvae of Crane Flies, which are found in grassland.

Out in the countryside in winter, European Starlings gather at dusk in huge winter roosts in reedbeds or woodlands – the chattering noises of the birds fill the air with what sounds like a massive starling conversation. Overhead in flight, the large flocks wheel and turn in the air in shifting shapes and sudden swoops, providing an avian spectacle that is called a 'murmuration'. To observe a large European Starling murmuration is a visual feast, which may draw admiring noises from the watching crowds, but, aside from the beating of thousands – sometimes tens or even hundreds of thousands – of wings, these flying flocks are silent; it's only when they find a perch that the murmuring breaks out.

The song of the European Starling at first does not seem exceptional, but it is delivered as a performance. The bird looks, and sometimes sounds, like a clockwork toy, perched in the open and giving its all. But listen carefully and you will hear some interesting passages. European Starlings are great mimics, and include snatches of other birds' songs in their own. You may hear the song of a Curlew, an open country bird that does not nest in towns, as you walk down the street, only to look up and find that a European Starling is responsible.

Starlings can also be taught to mimic human speech. Pliny the Elder claimed that it was possible to teach captive European Starlings whole

ABOVE
A glossy European Starling singing from a perch in a wood or town is a common sight across much of Europe. Its song may contain mimicry of songs of birds that nest hundreds of kilometres away.

sentences in Latin or Greek. The Welsh story, the Mabinogion (a collection written in the twelfth century containing even older stories), tells of Branwen (unhappily married to the King of Ireland) sending a European Starling from Ireland back to her brothers in Wales to ask them (in Welsh) to come and rescue her.

Shakespeare has Hotspur, in *Henry IV*, Part 1, planning to train a European Starling to keep mentioning the name of his brother-in-law, Mortimer, to the king in order to pester him to pay a ransom to free the captive man: 'The king forbade my tongue to speak of Mortimer. But I will find him when he is asleep, and in his ear I'll holler "Mortimer!" Nay I'll have a starling shall be taught to speak nothing but Mortimer, and give it to him to keep his anger still in motion.'

This single short mention is responsible for the arrival of the European Starling in North America, as the American Acclimatization Society set itself the goal of introducing to the USA every bird species mentioned in Shakespeare. It was sufficient reason for hundreds of European Starlings to be released in New York in the 1890s, since when the birds have spread to every state of the USA (including Hawaii and Alaska). They are now commonplace sights on a continent to which they were introduced and which they have made their home, in much the same way as European settlers did after they arrived on the *Mayflower* on the east coast of America. Americans and Canadians have mixed feelings about the European Starlings in their towns and gardens, some regarding them as a pest, but as birdwatcher Jeffrey Rosen put it in a 2007 *New York Times* article, 'It isn't their fault that they treated an open continent much as we ourselves did.'

The European Starling may not be the prettiest bird in the world (although it is very attractive when looked at in detail), or the most tuneful songster (although it has its moments), but it is tough, adaptable and successful. It's a bird that shares our cities with us, as well as our woods and fields, and it's a bird that we have taken with us on our travels – and may take for granted – and we would certainly miss its company if it disappeared.

LYREBIRDS

Menura novaehollandiae
& Menura alberti

Two species make up the entire family of lyrebirds: Superb and Albert's, named in honour of Queen Victoria's consort. They are found only in Australia (and feature on the Australian 10 cent coin), where they inhabit the eastern rainforests, feeding on insects, invertebrates and even small frogs turned up by scratching among the leaf litter with their powerful legs and feet. The male Superb Lyrebird is up to 1 m (around 3 ft) long, and gives its name to the family because of its extravagant tail feathers, including outer ones that resemble the musical instrument, the lyre. Albert's Lyrebird is smaller, and slightly less showy. In fact the Superb's feathers are not quite as 'lyre-like' as was first thought: when a bird was brought back to London in the nineteenth century it was reconstructed, wrongly, with the tail erect like a peacock's in display.

Male lyrebirds are territorial and their territories may encompass the home ranges of several females. Each male lyrebird has a display mound, an area of bare soil surrounded by thick vegetation in the case of the Superb, and trampled twigs in Albert's, from which he sings in order to attract females for mating. Males may sing for half the daylight hours in the hope of success. Perhaps the length of time spent singing demonstrates both the quality of the individual male and also that of the territory he has been able to establish and defend from other males. The male uses his extremely long, lyre-shaped tail, which he fans out and bends forwards over his body, to display to females who visit his mound to watch and listen. Having mated, the female alone builds the nest on or near the ground, lays usually a single egg and rears the chick.

Not content, it seems, with such spectacular displays of plumage, and perhaps to broadcast their message further afield, male lyrebirds have another remarkable talent. The syrinx (sound-producing organ) of the lyrebird is more muscular and more developed than in any other bird and can even produce two tunes at the same time. An example of this was a pet lyrebird which learned to mimic the flute-playing of its keeper before it was released into the New

OPPOSITE
The spectacular tail of the male Superb Lyrebird bears a striking resemblance to the ancient Greek stringed instrument from which the bird takes its name.

RIGHT
Lyrebirds are accomplished singers and mimics in the forests of eastern Australia. The male lyrebird has a more muscular and developed syrinx (sound-producing organ) than any other bird, allowing it to sing two tunes simultaneously.

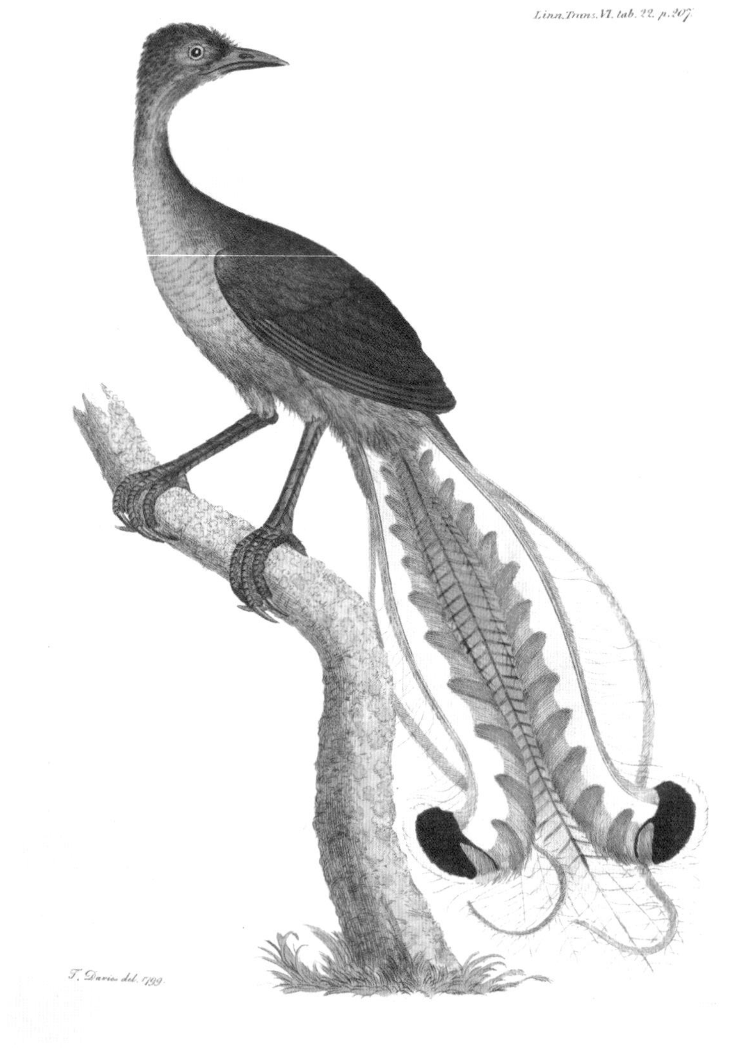

England National Park in New South Wales in the 1930s. In 1969, a recording was made of a lyrebird singing with flute-like tones in the same national park, and once the recording was analysed, and the lyrebird's own song was filtered out, two popular tunes from the 1930s were clearly identifiable. This story also testifies to the long life – up to thirty years – and faithful mimicry of the bird.

One lyrebird has been filmed (for the 1998 television series *The Life of Birds*) mimicking modern man-made devices in its song, including fire alarms and camera shutters, and even, rather poignantly to the human observer, the sound of a chainsaw cutting down the habitat on which it depends.

NORTHERN MOCKINGBIRD

Mimus polyglottos

One of the most accomplished of American songbirds, the Northern Mockingbird sings continually through the summer days and even at night. The nocturnal singers are often unpaired males who are putting in the hours to attract a mate. Many a jet-lagged traveller has woken in the darkness in an American city to be spellbound by the cascade of song pouring through the window from a Northern Mockingbird perched on a rooftop.

Males are mimics (their scientific name means 'many-tongued mimic'), and add songs to their repertoire throughout their lives. Many individual Northern Mockingbirds have over 100 songs that they sing, and some reach 200. As a species, Northern Mockingbirds have their own song that only they sing, but they add, in mimicry, those of over 30 other American songbirds, as well as those of frogs and crickets, the sound of barking dogs and also sometimes mechanical noises from the human world, such as car alarms, creaking gates and phone ringtones. This songster has both variety and stamina. Northern Mockingbirds will sing when in captivity and were prized cage birds in the nineteenth century, with particularly good songsters changing hands for up to $50, a large sum then.

There are now more Northern Mockingbirds nesting in urban than rural environments, and so this slender, pale grey-brown bird is a species that is familiar to many Americans. And there is evidence that individual Americans can be recognized by individual Northern Mockingbirds. In experiments, mockingbirds reacted with defensive calls to humans who had previously approached and touched their nests, but not to those who had approached the nest without touching it.

Female Northern Mockingbirds sing, too, although their song is most frequently heard during the autumn months when the sexes often hold separate territories. Males also have a distinct autumn song (September to November), which is different from their spring song when they are paired with females and

RIGHT
This early eighteenth-century portrait of the Northern Mockingbird by Mark Catesby, although inaccurately depicting its plumage, succeeds in capturing the character of this perky bird.

OPPOSITE
A group of Northern Mockingbirds attempt to drive off a rattlesnake that endangers a nest. The white wing-patches and white outer tail feathers are on display, and the scene is so well-captured that one can almost hear the birds' alarm calls.

raising a family together (February to August). The song is usually uttered from a prominent perch, and the bird's tail bobs in time with the song – as though the whole body is participating in singing.

Studies suggest that females are paying attention to the songs of their mate as well as to other neighbouring males through the breeding season. If a male flags in its singing, his mate may pay more attention to neighbouring rivals. Divorce, mate-switching and occasional mating with neighbouring males have been observed, and may be triggered by the quality of the songs – perhaps an indication of the overall health and well-being – of the males.

The Northern Mockingbird is found in all of the lower continental 48 states of the USA, but is perhaps most loved in the South, being the state bird of Arkansas, Florida, Mississippi, Tennessee and Texas. It is not, however, the chosen bird of another southern state, Alabama, the home state of author Harper Lee whose *To Kill a Mockingbird* is set in the fictional Alabama town of Maycomb. In the novel, Atticus Finch tells his son Jem that 'it's a sin to kill a mockingbird'. When Jem's sister, Scout, asks about this, Miss Maudie, a family friend, explains it is because 'Mockingbirds don't do one thing for us but make music for us to enjoy. They don't eat up people's gardens, don't nest in corn-cribs, they don't do one thing but sing their hearts out for us.'

Birds of Prey

Osprey • Bat Hawk • Snail Kite • Eleonora's Falcon
Eurasian Kestrel • Harpy Eagle • Secretary Bird

ABOVE Eurasian Kestrel.

Birds of prey, the falcons, hawks, eagles, vultures and others of the world, have always attracted humankind's interest and admiration. Although many eat carrion – vultures of course specialize in this – they are usually thought of as skilled hunters and were held in high regard by our hunter-gatherer ancestors for their prowess. And the powerful and skilful flight of some made them attractive to falconers and respected by naturalists. Evolution by natural selection has produced a wide range of variations on the theme of a hunter with a sharp, hooked beak and curved talons. The perhaps surprising range of prey species and how they are captured are explored in this section.

Most birds described here use their flying skills to catch their prey by surprise, even in the air. For instance, the Bat Hawk exploits the brief abundance of bats leaving their daytime roosts in the dwindling light and must have many successes in a short time to survive, and to reproduce. The swift Eleonora's Falcon inhabits Mediterranean islands and captures small migrating birds on their journey south to Africa in autumn.

Others swoop down on their chosen meal, such as the Osprey, a fish-eater, which dives into the water to catch its prey. Eurasian Kestrels hang in the wind looking for signs of small voles moving in the vegetation below them, while the Snail Kite of North, Central and South America, specializes in small aquatic molluscs, which it can spot from a surprisingly great height.

Two of our other birds of prey use slightly different tactics. The powerful Harpy Eagle is a rainforest predator of monkeys and sloths, which it catches by stealthily moving from tree to tree. The Secretary Bird of the grassy plains of Africa is perhaps the most unusual in not using flight at all – it stalks on long legs and uses its feet to grab and subdue a wide range of prey.

These are simply examples of the extraordinary range of ecologies adopted by birds of prey. Their position at or near the top of the food chain makes them appear powerful, but it also makes them vulnerable. As hunters, often large ones, these birds need a sizeable range in which to find their prey. In the past, the accumulation of toxic chemicals has greatly reduced the populations of some. So their continued existence in an area is often seen as a sign of habitat quality. A world rich in birds of prey is likely to be rich in other species too, and arguably provides a healthy environment for us.

OSPREY

Pandion haliaetus

ABOVE
Ospreys lay two to four beautifully marked eggs, which are highly prized among egg-collectors, making them a target for theft. Some Osprey nests are protected by round-the-clock vigils or CCTV cameras, others by barbed wire. The tree in which the famous Loch Garten Ospreys nest, near Aviemore in the Scottish Highlands, has had most of its branches removed to prevent egg thieves from reaching the nest.

OPPOSITE
An Osprey catches a fish in a freshwater lake in North America, while two adult Bald Eagles – also frequent fish-eaters – circle above. The eagle closest to the Osprey looks as if it is set to steal the fish.

Relatively few birds of prey include live fish in their diet and none is as dependent on them as the Osprey. Fish make up 99 per cent of the diet of Ospreys, which they catch by diving through the air, feet first, into the water. Despite this specialization, or perhaps because of it, the Osprey is found throughout the world, occurring on every continent except Antarctica.

As well as being a widespread species, Ospreys are migrants, with North American birds wintering in South America, and European and some Asian birds migrating into Africa. For example, Ospreys nesting on Martha's Vineyard, Massachusetts, USA, winter in western Brazil, and Ospreys from northern Scotland travel to coastal Mauritania, Senegal and Guinea-Bissau in west Africa to winter. The autumn migration tends to be at a more leisurely pace than the return to their nesting sites by lakes and rivers in the spring. Most of the journey is travelled in daylight hours, although larger water bodies, often stretches of sea, are sometimes crossed at night. Around 260 km (160 miles) are travelled each day. Ospreys pair for life and birds often return to the same nests, patching and adding to them with more twigs each year, so that they can become quite large structures.

Ospreys spot their prey from circling high above the water and then plummet, talons first, into the water to grab the fish. Their victim is often around 30 cm (1 ft) in length, which, since the birds themselves are 54–58 cm (21–23 in.) tall, is quite a feat. The Osprey is unusual in having a reversible outer toe that allows it to grasp a fish with two toes on either side to reduce the chance of dropping the slippery prey, which may still be struggling to escape as the Osprey flies off with it. In addition, their feet have barbed pads which help to secure a firm grip. Ospreys carry their prey aligned with the direction of flight and head first to reduce wind resistance.

The Osprey's hunting technique is impressive and a joy to watch, and around a quarter of attempts result in the capture of a fish. Sometimes the

OSPREY, FEMALE.

Osprey will simply pluck a fish from close to the surface of the water with hardly a break in momentum. At other times, the plummeting bird, legs extended and talons stretched forward, splashes into the water at great speed and is fully immersed before it starts flapping to escape the water with or without a fish in its claws. If the Osprey does have a fish, particularly a large one, then just for a second or two, this graceful bird looks as if it is struggling to regain the air; its wing-beats appear heavy and laboured, until it emerges from the water and picks up speed. Once safely airborne again, the bird often shakes its head, back and wings to rid them of as much water as possible, which would otherwise weigh it down.

Along with other birds of prey, the Osprey is a top predator and was greatly affected by the use of pesticides, especially DDT (dichlorodiphenyltrichloroethane) from the 1940s until its use was generally banned in the 1970s. Although DDT was an agricultural insecticide, it still found its way through run-off into streams and rivers and into coastal waters. Ospreys nesting in areas of extensive DDT use, perhaps to kill mosquitoes, built up high levels of DDT in their bodies, which interfered with egg-shell formation and reduced the proportion of eggs that hatched. Some US Osprey populations were declining at a rate of 25 per cent per year at the height of the impact. Since the banning of DDT (and other organochlorine pesticides) Osprey numbers have recovered and expanded. They are now regularly seen in areas of high human populations as they habituate fairly readily to the presence of people and will nest on man-made platforms specially provided for them.

As Osprey populations continue to recover from the impacts of pesticide pollution, it is to be hoped that the magnificent sight of a fishing Osprey will become an increasingly common experience that more people will be able to enjoy.

OPPOSITE
This illustration clearly shows the large blue talons and long claws that are necessary to hold a struggling and heavy fish as the Osprey lifts off from the surface of the water and flies away with its prey.

BELOW
An Osprey eats its fish under the gaze of a White-tailed Eagle (as an immature one flies off). Both species were driven to extinction in the UK in the early twentieth century, but have been re-established through conservation efforts.

BAT HAWK

Macheiramphus alcinus

A specialist bird of prey, the Bat Hawk does indeed eat mainly bats, which it captures in flight with its feet, although it also catches some birds, too, and large insects. Bat Hawks are mostly dark brown or black, with a small crest, and their legs and talons are long and slender. The species is widely but thinly spread across sub-Saharan Africa and Indonesia, Malaysia and Papua New Guinea. Despite such a large range, the total population is thought to number, at most, only around 6,500 individuals.

ABOVE
Bat Hawks generally sit motionless in trees during the day, with most of their hunting taking place at dusk. This portrait shows the large eye necessary for crepuscular hunting, and the bird's long, thin legs and talons.

Bat Hawks depend on there being large bat colonies where they can reliably hunt. Some hunting takes place at dawn or around artificial lighting if this attracts bats, but most occurs in a 20-minute period at dusk, as the bats are emerging to feed. The Bat Hawks may take a bat every couple of minutes, and around half of attacks result in successful captures – successful for the Bat Hawk, not so successful for the bat. It's a brief period of abundant food, which occurs each day and which the Bat Hawk must exploit to the fullest extent. The birds weigh around 600 g (21 oz) and catch small bats, which are eaten whole, in flight, weighing 20–75 g (0.7–2.6 oz).

The Bat Hawks hunt by sight, which raises the question: why don't the bats come out a bit later when it's completely dark? The answer may be that the bats are emerging to exploit an insect food resource that peaks at dusk, and any delay may reduce their own feeding success. So the Bat Hawks are successful because the bats are in a hurry to feed themselves. Perhaps a better way of looking at it is that the bats emerge sufficiently late to avoid most other avian predators; only the Bat Hawks have evolved the ability to exploit this specialized prey resource.

SNAIL KITE

Rostrhamus sociabilis

ABOVE
A Snail Kite in its wetland habitat. The large curved beak has enabled the kite to extract the flesh of two freshwater snails from their shells; these form almost the entire diet of this specialized hunter.

The prey of the Snail Kite is not so fast moving as that of some other birds of prey. This bird eats snails, freshwater snails, and almost nothing else. A medium-sized bird of prey, the Snail Kite uses its long, slim and deeply curved beak essentially as a tool for winkling the snails out of their shells. One of its favourite molluscs is the apple snail (about the size of a ping-pong ball). This has various defences against other predators, but relies on crypsis – trying to avoid detection – against Snail Kite predation.

The Snail Kites fly surprisingly high over wetlands when hunting – their eyesight must be very good to spot their small prey. Snails are plucked from the vegetation and from up to 16 cm (just over 6 in.) below the water surface in the bird's feet and then taken to a perch. With the snail held firmly in the talons, the curved beak gets to work, neatly snipping the columellar muscle which attaches the snail to the inside of its shell.

This almost exclusive diet of freshwater snails limits the habitats where the Snail Kite can be found – lakes, ponds, ditches and flooded fields – and it can be seen in flocks in such places. But the overall range of the Snail Kite is quite wide. It occurs in South America and patchily in Central America, with a few locations in Florida, USA, and the West Indies. In its Florida range there is a non-native species of apple snail which is much larger than the native species. Snail Kites are unsurprisingly attracted to this larger snail, but are much more likely to drop it (44 per cent of the time) than they are the native snails with which they are more familiar (dropped only around 1 per cent of the time).

ELEONORA'S FALCON

Falco eleonorae

The Eleonora after whom this falcon is named was Eleonor of Arborea, a fourteenth-century warrior ruler of a third of Sardinia who established a body of laws that were highly regarded and survived as the basis of Sardinian law into the early nineteenth century. Eleonor was also interested in birds, and legislated to protect the bird of prey that now bears her name.

This dark, medium-sized migratory falcon with long wings and slender body spends its winters in Madagascar. It then returns to its breeding colonies, which are primarily located on cliffs on Mediterranean islands, arriving in late April. Two thirds of the world populations inhabit Greek islands, but large numbers are also found in the Balearic Islands of Spain, and outside the Mediterranean in the Canary Islands and the Atlantic coast of Morocco.

ABOVE
This mid-nineteenth-century portrait of an Eleonora's Falcon, while perhaps not conveying the bird's talents as a powerful killer, does capture its plumage to great effect.

OPPOSITE
An early twentieth-century depiction of an Eleonora's Falcon shows both the pale and dark plumage types, as well as hinting at the social behaviour and Mediterranean coastal habitat of this species.

The timing of the breeding season of most bird species is usually connected to the annual peak in food availability. In high latitudes the long summer days promote increased plant growth, which is the basis for a flush of animal life too. Different species have evolved different strategies for cashing in on this peak in living biomass, but the upshot is that in high latitudes most produce their young in the spring and early summer months. In the tropics, where day lengths and temperatures vary little, it is the seasonal rains that often determine the breeding seasons of species. And so the Eleonora's Falcon, which nests and raises its chicks in the later summer and early autumn months of July–October, at first seems an anomaly.

The explanation for the apparently delayed breeding season of this falcon can be found in the mass southward migration of small songbirds across the Mediterranean Sea as they return to their wintering grounds in Africa. The

2
1

hatching of the falcon eggs in late August and through September is thus perfectly timed to coincide with the huge numbers of these migrating birds, many of them youngsters produced over the previous summer months, which now struggle across the sea towards Africa. The sheer abundance of migrant birds, a large proportion of them naive and tired by their journeys, makes them a rich resource of weak and vulnerable prey, which the Eleonora's Falcons exploit for feeding their chicks.

ABOVE
The breeding season of the Eleonora's Falcon coincides with the mass migration of small songbirds moving southwards across the Mediterranean Sea – many of which are young and tired from their travels, making them easy prey. This provides a valuable food source for feeding the young falcons, two of which are seen here.

The Eleonora's Falcons nest on coastal cliffs and detect their prey as they approach landfall, flying out to capture them above the sea. The attacked prey can only seek shelter by reaching land or by outmanoeuvring the falcon until it stops giving chase. Eleonora's Falcons are fast and agile predators, and can climb very rapidly for their size – Swifts are one of the few species that can regularly escape a chasing Eleonora's Falcon by climbing higher. More often the intended prey attempts to escape by diving steeply and then pulling out of the dive sharply, or by a quick turn in level flight. These aerial manoeuvres are played out above the sea, with the prey attempting to escape the falcon in the short term and then to reach the relative safety of land.

Many species of birds are taken as prey; over 100 have been recorded – warblers, shrikes, chats, larks and pipits are all included in the diet. Few falcons nest colonially, but in the case of the Eleonora's Falcon there is little possibility of defending a food resource territorially since it arrives through the air almost like plankton. Nest-sites are abundant on the cliffs, and colonies of up to 300 pairs exist. Out of the breeding season, the main prey for adults consists of large insects.

The autumnal nesting of the Eleonora's Falcon, which at first might seem odd, on closer examination appears very well adapted to take advantage of the seasonal abundance of migrating birds. And the attacks by this falcon add to the dangers of migration for small birds making their huge journeys between the European breeding grounds and African wintering areas.

EURASIAN KESTREL

Falco tinnunculus

ABOVE
Kestrel chicks in a nest await the delivery of their next meal, most likely a vole or a mouse. There may be up to six chicks to feed.

The Eurasian or Common Kestrel is a familiar bird of the Eurasian open countryside – familiar both because it is widespread and numerous, and also because it often searches for its prey by hanging in the wind in hovering flight, which makes it readily identifiable as it scans the ground below for its preferred food of small voles. One alternative name for the species is Windhover. In a sonnet of the same name, Gerard Manley Hopkins admires the Kestrel's ability to master the air like a horseman masters his steed. The poem, published in 1918, long after Hopkins's death, was regarded by the author as his best work.

A hovering Kestrel, perhaps only briefly seen at the roadside as the observer rushes past at speed in a vehicle, makes a characteristic 'Y' shape, with the spread tail forming the stem of the 'Y' and the outstretched wings the two arms. The Kestrel faces into the wind, and essentially flies into it at the same pace as the air current blows it backwards in order to retain position. Often its wings seem held still, and angled so that the wind current provides the uplift needed to keep the bird steady at a constant height. A closer look at the bird, though, will reveal that the wing and tail feathers are in constant motion, adjusting to the gusts of wind. Quick flaps of the wings prevent the bird from being pushed backwards. What looks effortless from a distance is shown to be a skilful command of the air at close view, and Hopkins was right to celebrate 'the mastery of the thing!'

Observations of Kestrels in the wild have discovered that they mostly hover in moderate wind speeds of 21–42 kilometres per hour (12–25 mph). Calculations have shown that these are the wind speeds that minimize the energetic costs of hovering flight: at slower speeds, the Kestrel finds it too difficult to hang above the

Tinnunculus alaudarius. Briss

ground (the updraft delivered by its wings is reduced) and at faster wind speeds the bird has to use much more energy not to lose ground. So if the wind is just a faint breeze or it becomes too strong, you will notice Kestrels spending much more of their time perched in trees and on posts scanning the area below them.

When you see a hovering Kestrel, it is almost always looking down at the ground with its sharp eyes, searching for food, primarily its favourite small voles but also young birds, large insects such as beetles and sometimes reptiles including lizards. Voles make up a very large proportion of its prey, and it is thought that the Kestrel uses the reflection of sunlight off vole urine on the grass, which reflects in the ultraviolet light frequencies (invisible to human eyes), to spot areas where voles are numerous, thus aiding more efficient hunting. The adults each consume around six voles each day, but when they have nestlings to feed, they must find an extra three to four voles each day for each chick – and there may be six chicks for the two parents to feed, thus approximately trebling the number of voles that the parents must catch. The young remain in the nest for four to five weeks and then need to learn how to hunt for themselves.

Because it catches almost all of its prey by pouncing on it on the ground rather than in dramatic fashion in aerial chases, the Kestrel was not one of the birds of prey particularly prized in falconry ('a Kestrel for a Knave or Servant' decreed the fifteenth-century *Book of St Albans*). And yet its hovering flight and keen eyes represent a different set of finely evolved skills, ones at which we can marvel as we see the characteristic shape of a hovering Kestrel hanging apparently motionless in the air while we rush past in our busy lives. We should pause to acknowledge that the Kestrel's life is a very busy one too.

OPPOSITE
A male Kestrel (the female lacks the blue head and has a barred back) at a nest with three chicks of very different ages, which illustrate the transition from downy new chick to the feathered youngster almost ready to attempt flight. Nests can be in buildings, on cliffs, in holes in trees or in old Carrion Crow (*Corvus corone*) nests.

BELOW
Another male Kestrel, surveying a dramatic landscape perhaps more typical of a Peregrine Falcon (*Falco peregrinus*).

HARPY EAGLE

Harpia harpyja

In ancient Greek myth the harpies were legendary bird-monsters with human faces. According to earlier versions, their female faces were generally beautiful, but later, in Roman and Byzantine times, they were loathsomely ugly. In the case of the Harpy Eagle of Central and South America, which is named after them, the face is certainly commanding, with its circular arrangement of feathers and piercing eyes, dominated by a massive bill. The head is decorated with a striking crest of feathers of different lengths, which can be raised, adding to the bird's rather menacing appearance.

The Harpy is in fact one of the world's largest eagles – approximately 90–105 cm (35–41 in.) tall and with a wingspan up to 2 m (6½ ft) – and is one of the most powerful-looking of birds. As with most birds of prey, the female is larger than the male; in this case the disparity is extreme, with the female around twice the weight of her mate.

The mythical harpies stole food and carried away murderers to the vengeful Furies. Harpy Eagles also carry off their victims, usually medium-sized mammals caught within the upper canopy of lowland rainforests. The rear talons of female Harpy Eagles will be as big as a Brown Bear's claws and are designed to grab, kill and carry mammals as large as Howler Monkeys (8 kg, around 18 lb) and Brown-throated Sloths (6 kg or 13 lb). The Harpy Eagle hunts by moving stealthily through the forest from tree to tree, to look and listen for its prey, which also includes other mammals, iguanas and birds. Hearing seems quite important in prey detection and the Harpy Eagle can raise and lower the feathers of its face to help concentrate sounds to its ears.

But even such impressive physical power cannot protect the Harpy Eagle from its biggest threat – habitat loss. Large avian predators, just like large mammalian predators such as wolves, need extensive areas in which to hunt, and habitat destruction has reduced their numbers, particularly in Central America. The adults are long-lived, but the population is vulnerable to another

OPPOSITE
A Harpy Eagle looks straight at us with its piercing eyes, while clasping a Scarlet Macaw (*Ara macao*) that it has caught. The powerful nature of this bird is expertly captured by the artist.

threat: adults being killed for their feathers, which have been used in shamanic rituals. In one case, villagers killed a Harpy Eagle simply because it was so large and fierce-looking a predator that they feared it might take children.

Harpy Eagles do not nest until they are five years old, and even then only do so every two to three years. When nesting successfully, they usually only raise one chick from the two eggs they lay. The nests are found in the tallest trees of the rainforest, often Kapok trees, and are big structures made from hundreds of branches, built near the top of the trees at a height of 40 m (130 ft). From this solid platform, with its commanding view, Harpy Eagles look down on the forest and fully merit their Brazilian name of Royal Hawk (*gavião-real*).

RIGHT
Despite being a consummate predator and exuding an aura of power, the Harpy Eagle is vulnerable to rainforest destruction. A decline in numbers is driven by the pressure on its habitat by an ever-growing human population that needs land for growing food.

SECRETARY BIRD

Sagittarius serpentarius

ABOVE
A fierce-looking Secretary Bird towers over a lifeless snake. Native to sub-Saharan Africa, the Secretary Bird uses its long, strong legs to stamp on and kill prey such as snakes.

The Secretary Bird gets its unusual name from the spiky crest of individual feathers arrayed around the back of its head, which is thought to be reminiscent of the bunches of quill pens secretaries in former times kept tucked behind their ears. These distinctive plumes are black, in contrast to the bird's generally white and grey head and body, and are slightly longer in males than females. Secretary Birds live in the grasslands and savannah of sub-Saharan Africa, and although they can fly, these large birds, around 1.2 m (4 ft) tall, with their extremely long legs – the longest of any bird of prey – are mostly terrestrial and catch the majority of their prey on foot.

Not such specialist eaters as others in this section, Secretary Birds consume a wide range of prey up to the size of a hare, including birds, reptiles and large insects. Snakes were once thought to be the bird's main prey item (as hinted at in its Latin name, *serpentarius*), but they simply form part of its very varied diet. However, the Secretary Bird's ability to kill snakes, including poisonous ones such as puff adders and cobras, has won it admiration in many African countries and it is sometimes kept as a garden pet because of its snake-killing abilities. It is the national bird of Sudan, as well as featuring in the coat of arms of South Africa.

The Secretary Bird stalks on its tall legs – the upper part covered in black feathers, the lower part with scales, perhaps for protection – taking great strides through the grass as it hunts for its prey. It can also use its legs and feet to tread on bushes to flush out anything it can catch, and will even wait near grassland fires to attempt to pounce on potential prey fleeing from the flames. Individuals or pairs, or sometimes small groups, can be seen stalking through the grassland. When a prey item is spotted the Secretary Bird may chase it

R P Nodder Del et Sculpt

through the grass, running at speed with its wings open, before stomping on its victim with its powerful legs and feet to subdue and kill it. It is said, though it seems unlikely, that a stomp from a Secretary Bird can break a human arm; it can certainly kill a snake or rodent. Once killed, prey items are swallowed whole rather than being dismembered and eaten piecemeal. Having such long legs does have one disadvantage: the Secretary Bird must crouch in order to drink, as its neck will not reach much further than half way down its legs.

Secretary Birds nest in the tops of small trees, often acacia, or even smaller bushes. The nest, made of sticks and lined with grass, is built by both members of the pair. As they use the same nest in successive years it can become large, often almost 3 m (10 ft) across. The clutch consists of two or three off-white eggs, which are incubated for around six weeks. The adults regurgitate food for the young chicks once hatched and then start bringing whole prey to them as they grow bigger. The adults roost in trees overnight, often at the nest-site. They are thought to pair for life, and the members of a pair are often found feeding together. The young take their first flights at 10–15 weeks of age and then accompany their parents stomping through the grasslands in their search for prey.

Because it is such a large bird, becoming airborne usually takes a long run-up through the grass to gain speed. Once in flight, however, the Secretary Bird is a good flyer and resembles a stork or crane because of its long, extended legs, which are matched in size by two elongated tail feathers. Undulating display flights are part of the courtship ritual and are accompanied by guttural calls; at other times these birds are largely silent.

Most birds of prey are known and admired for their flying ability, but the Secretary Bird is largely terrestrial, and a good walker, covering long distances every day – possibly as much as 32 km (20 miles).

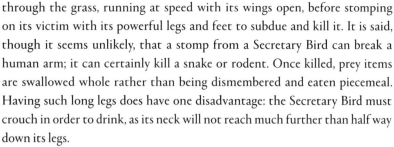

OPPOSITE
The Secretary Bird's long legs carry it through the grasslands of Africa, where it finds most of its prey. At pace it leans forward and takes large steps, while its long tail provides a counterweight for balance.

BELOW
Although they eat a range of prey, Secretary Birds are famed for their ability to kill snakes, including venomous ones.

OVERLEAF
The Secretary Bird takes its name from the plumes on its head, which were thought to resemble a secretary's quills tucked behind the ear.

Secretary Bird-
Sagittarius
Serpentarius.
Dangela
Gojam-
Mar. 25. 1927.

Feathered Travellers

Common Swift • Bar-tailed Godwit • Cuckoo • Arctic Tern
Hummingbirds • Peregrine Falcon

ABOVE Hummingbird.

The dinosaur ancestors of today's birds evolved feathers as a means of insulation, and they are still excellent insulators for both birds and us. But feathers also opened up the skies to birds – first perhaps through gliding from tree to tree and then in flapping flight. These days, only about 40 of the world's 10,000 species of birds are flightless (including the Ostrich, Emu and penguins); the rest retain an ability that humans have long envied.

Birds can fly unhindered over obstacles that earthbound humans have to climb over or circumvent. Only in the last two centuries has humankind been able to take to the skies, but even then we are strapped into contraptions as the birds swoop, glide and hover apparently at will.

Of all the birds in the world, the Peregrine Falcon is the fastest, and it uses its speed to catch its prey – almost entirely other birds in flight. Its prowess in the air has made it a favourite of writers, falconers and naturalists. Common Swifts are very proficient flyers, doing almost everything on the wing: they feed, mate and can even sleep in flight. Their young remain continuously airborne for two years before nesting themselves.

Other birds excel in terms of the distances they fly. Arctic Terns nest in high latitudes in colonies. Their chicks will migrate to the Antarctic Ocean before returning to the Arctic to nest themselves. The Bar-tailed Godwit is also a long-distance traveller, whose prodigious journeys are often made in non-stop flights covering thousands of kilometres.

The return of migrant species is often a welcome sign of the changing seasons. In northern latitudes the arrival of the Cuckoo marks the beginning of spring. We are now learning more about how far these birds travel and how they spend their time during the non-breeding season. Perhaps most surprising, some of the tiny hummingbirds, the smallest birds, which hover with remarkable precision to feed on the nectar inside flowers, are also long-distance migrants.

Ex Collectione plurimum reverendi Domini Adriani Buurt.

COMMON SWIFT

Apus apus

Swifts are a sure sign of summer. One of the last migrants to arrive back in their northern European breeding grounds, after the warblers, swallows and cuckoos, they are also one of the first to leave. They stay to nest for a brief few months, from May to mid-August, and then are gone. Other migrant birds may arrive on the coat-tails of winter and leave when autumnal colours begin to spread across woodland, but the Swifts are birds of summer.

While they are with us, the skies above towns and villages are graced with the black, scythe-shaped outlines of these consummate flyers. Sometimes the air fills with their distinctive calls, especially in June and July when they form what are aptly known as screaming parties, as a group of birds swoops at great speed. Their screeches gave them the name of Devil Bird in some languages.

These blackish-brown birds, with their short, forked tail, feed on the wing, roaming the skies in search of insects. But they also mate on the wing and normally even sleep on the wing. It's as though they disdain the earth below and only deign to touch it when they must – and that's solely to nest.

Swifts nest almost entirely in buildings, usually in small gaps in the roofs of old buildings; a structure with nesting Swifts has almost certainly stood for centuries. But this poses the question of where Swifts nested before we provided them with house-space. The answer is they nested in trees – and they still do in small numbers in ancient forests. To find tree-nesting swifts you must penetrate such places and look for old woodpecker nests in which the Swifts will lay their two or three white eggs. But only a tiny proportion of Swifts now nest in natural nest-sites, probably less than 1 per cent – the rest live with us in cracks and crevices that we accidentally leave for them, although increasingly Swift nest-boxes and nest-bricks are being included in new buildings to provide nesting sites. They even gather the material for their nests on the wing, and bind it together with their saliva, like swallows (though they are not closely related). Swifts enter their nests at top speed – blink and you'll miss them!

OPPOSITE
Common Swifts nest mainly in the roofs of buildings, although a small number can still be found in natural nest-sites in old trees. They pair for life, with the same nest-sites used by the same pairs or surviving individuals each year. Sometimes pairs fight over ownership of a nest-hole.

So when the young leave the nest, just a very few now fly through ancient trees where Swifts must have previously flown for thousands of years. Most launch themselves into the air over suburban gardens and even cities. Once in the air, a young swift will stay on the wing, without landing, for almost two years until it seeks a mate and its own nest-hole in which to rear a family. For those two years it spends every moment on the wing, travelling to Africa to feed above tropical rainforests and open savannahs. What about sleeping? Swifts, like many other birds we believe, are able to close down one half of their brain in sleep and use the other half to continue flying on 'auto-pilot'. What would our lives be like if we could do something similar?

Adult Swifts departing from Britain in late July travel through France and Spain, down into Morocco and then continue south through West Africa. By 1 August they may be in Senegal or Mauritania; northern Europe, which they left a mere ten days before, is still basking in full summer 6,400 km (4,000 miles) away. The Swifts then travel west to the Democratic Republic of the Congo (DRC), where they feed above the rainforests until December. They then head further east to the coast of Mozambique for a short while, before returning to the DRC until early April (when many other migrant birds are already back in their European breeding grounds). Moving back through West Africa, Swifts stop for a while in Liberian skies, cross the Sahara and return to northern Europe in early May, to become, again, a very welcome feature – though sadly in declining numbers – of the summer scene above European towns and cities.

ABOVE
After leaving the nest where it was raised, a Common Swift will spend its first two years in constant flight, without landing, until it is time to find a mate and its own nest-site. It will sleep on the wing and travel from Europe to Africa, spending much of the northern winter hunting insects over the rainforests of central Africa.

OPPOSITE
There are around 100 species of Swift, all powerful and gifted flyers. Over their lifetime many of the migratory species will fly millions of kilometres, and some can reach top speeds of over 170 kph (105 mph).

Allan Brooks.

BAR-TAILED GODWIT

Limosa lapponica

ABOVE
Bar-tailed Godwits double their weight before setting off on their long migration from the Arctic to the southern hemisphere using their long beaks to probe for food. They may not look like powerful flyers, but they make a non-stop epic journey from Alaska to New Zealand of 11,000 km (6,800 miles).

A coastal wading bird that may not otherwise appear particularly remarkable, the Bar-tailed Godwit makes an astonishing eight-day, non-stop flight on its autumn migration from Alaska to New Zealand, covering 11,000 km (over 6,800 miles) without rest, food or water. Flying at a height of 3–4 km (around 2–2½ miles), seeking tailwinds, the godwits achieve an average speed, day after day and night after night, of 56 kph (35 mph).

The details of this migration were revealed through technology that sends details of the birds' location and activity to scientists via satellite. A female godwit, equipped with a satellite tag in the northern winter, flew directly from the feeding grounds in New Zealand to a wetland on the North Korea–China border, a distance of 10,200 km (6,340 miles), stayed there for a few weeks and then continued her journey, adding another 5,000 km (3,107 miles), to Alaska to breed. It was this same bird that made the non-stop journey back to New Zealand by the more direct route of flying across the Pacific Ocean.

To prepare for their journey, the godwits feed for days, by probing, with their long, slightly upcurved beaks, in soft soils and coastal sediments for invertebrates such as worms and molluscs. They double their body weight with stored fat. Their internal organs, especially their guts, shrink to reduce weight and to provide more room for all that fat.

It is thought that the Māori may have discovered New Zealand by following migrating birds such as godwits.

CUCULUS canorus. *Gemeiner Kuckuk.*
1. M. 2. W.

ABOVE
The song of the Cuckoo is a sign that spring has arrived. Cuckoos lay their eggs in the nests of other birds, and their arrival in Europe in late April coincides with the start of the egg-laying period for most birds. Cuckoos, free of any parental duties as their young are tended by the host species in whose nests they lay their eggs, set off on a southward migration back to Africa in early July.

CUCKOO

Cuculus canorus

A very easily recognized and remembered song, the two-note 'Cuck–oo' has heralded the true arrival of spring for centuries in Europe, as the male makes his call, signalling that this migrant bird has returned. But although many may know the song of the Cuckoo, probably far fewer would recognize the thin, streamlined, blueish-grey bird if it were to fly past. When perched, often as they sings, Cuckoos hold their wings drooped down from the body in a pose that looks a little like that of a bird with injured wings.

Famously, the Cuckoo is a brood parasite – females lay their eggs in the nests of other birds while they're unguarded, removing one of the host's eggs and replacing it with their own. That's a remarkable enough story in its own right, but it means that Cuckoos have no duties of parental care, leaving the feeding and rearing of their young to the unsuspecting host bird, sometimes much smaller than the well-fed and voracious Cuckoo chick. Once the opportunity to lay more eggs is finished, the adult Cuckoos can move on to wherever the food is most abundant and easiest to find for the rest of the year.

Cuckoos caught by scientists have been fitted with devices that transmit their locations in real time, via satellite links. This has disclosed the details of their migration in fascinating detail. As in the traditional verse 'Cuckoo comes in April; Sings his song in May', Cuckoos do arrive on their breeding grounds across northern Europe in April and May and then depart, unencumbered by family duties, as early as the first few days of July. By early August, when young Cuckoos can still be seen in the European countryside, many of the adults have already made their way through Europe, crossed the Mediterranean Sea, spent a little time in North Africa and flown across the Sahara Desert and arrived

Fr. Naumann pinx & sculpt 24.

RIGHT
This young Cuckoo is being raised by a much smaller Meadow Pipit (*Anthus pratensis*), a common host species on moorlands and heaths. The 'offspring' and 'parent' have different ways of coping with the winter: the young Cuckoo will migrate to Africa, while the Meadow Pipit will remain in Europe, seeking milder conditions in the lowlands or on the coast.

OPPOSITE
Studies using satellite tags suggest that individual adult Cuckoos (above) tend to use very similar migration routes year after year. Young Cuckoos (below) have to find their own routes, with no parental help, as they set off on their journeys to distant Africa.

in the Sahel region. There they spend some time resting and feeding, but in mid-winter the Cuckoos are on the move again, to central Africa, in areas of rainforest in the Democratic Republic of the Congo, Central African Republic and Angola.

The information gained from Cuckoos equipped with satellite tags in the UK, Germany and Belarus has opened our eyes to the complexity of this migration. We have learnt that when conditions for migration are unfavourable, birds will back-track hundreds of kilometres to wait until things improve for onward travel. Some individuals have followed very similar migration paths in successive years, and yet the overall variety of migration routes is broad. The return journey is often by a completely different route from the one to Africa.

And we now know about the whereabouts, day by day, of a few dozen individual Cuckoos in incredible detail. By the time a Cuckoo returns to Europe he may have visited 30 countries, passing the cities of Rome, Tripoli, Kinshasa, Lagos, Timbuktu and Madrid. He will have flown over the Mediterranean Sea and across the Sahara Desert and then the great central African rainforests. The simple two-note song of the Cuckoo has always been a familiar reminder to us that spring has arrived and summer is on its way, but this represents just one brief moment in the much more complex lifecycle and behaviour of this bird, and few people perhaps realize how far it travels and how many countries it visits before it comes 'home' to us.

ARCTIC TERN

Sterna paradisaea

A global traveller, the Arctic Tern lives in perpetual summer and often in almost constant daylight. In the long days of the northern summer these terns nest in mostly Arctic latitudes, and then 'winter' in the Antarctic regions in the southern summer. Arctic Terns nest on the ground in large colonies, with three eggs laid in simple scrapes in the vegetation or sand. The parents feed their young on fish caught by diving, wings folded back, headfirst and, more importantly, beak-first, into the waves. This plunge-diving method restricts Arctic Terns to feeding on fish shoals within around a metre of the sea surface. The terns gorge on shoals of sandeels, herring, capelin and sometimes cod.

It is this great productivity of the far northern coastal waters that attracts the terns. It's a short period of plenty, but one that will be repeated in the Antarctic Ocean in six months' time – so that's where the Arctic Terns then travel to 'winter'. We now know the routes taken by Arctic Terns on their annual round trip of 70,000 km (43,500 miles). On their southward migration they stop for a month or so in the seas about 1,600 km (995 miles) north of the Azores – probably because this is the last oceanic area of high food abundance before the Antarctic. They then fly down the coast of West Africa to the Cape Verde Islands; some birds continue down the African coast, while others cross the Atlantic and travel down the coast of Brazil and Argentina. All are heading to the Weddell Sea, where they arrive in time to exploit the flush of marine life.

As the southern summer comes to an end it is time for the terns to head back north again to the Arctic, though this time they take a different route. Rather than hugging the coast of Africa, they cross the Atlantic and pass along the shores of northern Brazil, Guyana and Venezuela, skirting the Caribbean and crossing the west Atlantic back to Greenland. This route adds thousands of kilometres to their journey and at first puzzled scientists, but the explanation appears to be in the prevailing winds. Although the route is further, it benefits from more tailwinds that make the journey much easier.

OPPOSITE
The Arctic Tern has crimson legs and beak and a black cap. It catches small fish such as sprats by diving headfirst into the sea, snatching its prey in its sharp beak. Arctic Terns also defend their nests fiercely and may attack anyone who enters their nesting colonies.

1	2	3	4
Topas-Colibri.	Gemeiner Colibri.	Langsdorfischer Colibri.	Geschilderter Colibri.
Trochilus pella.	Trochilus colubris.	Trochilus Langsdorfii	Trochilus scutatus.
Colibri topaze.	Oiseau-mouche rubis.	Oiseau-mouche Langsdorf.	Oiseau-mouche écussonne.

HUMMINGBIRDS

Trochilidae

All of the 340-odd species of hummingbird live in the Americas; they can be found from southern Alaska to Tierra del Fuego, though half the species live in the Amazon basin. Hummingbirds are often tiny – the smallest species is also the world's smallest bird, the Bee Hummingbird, which lives in forests on Cuba and weighs less than 2 g (0.07 oz). They are often very brightly coloured, and hover almost in a blur, like tiny jewels. Their English name, hummingbird, derives from the sound their rapidly beating wings make – in normal flight hummingbirds beat their wings around 50 times per second.

Such energetic flight requires large amounts of high-octane fuel, and most hummingbirds feed on the calorie-rich nectar of flowering plants. They will consume their own body weight in nectar each hour, feeding several times an hour, for 12 hours a day. Most hummingbirds feed by hovering and flying precisely forward so that their long bills, some with special shapes adapted to particular flowers, enter the nectaries of the plants, allowing their elongated tongues to suck up the energy-rich liquid. The hummingbird then reverses backwards in flight and moves on to the next flower. They are the only birds that can fly backwards as well as forwards. The precision with which all this is accomplished appears almost effortless, but the hummingbird's minute heart is beating at 20 beats per second.

In addition to such amazing feats of nimble flying, those hummingbirds living at high latitudes, such as the Rufous Hummingbird, which nests in Alaska, and the Green-backed Firecrown of southern Argentina and Chile, are

ABOVE
In flight, the wings of hummingbirds are a blur and make the whirring sound that gives the bird its name. It is a challenge for any artist to capture this aspect of their behaviour, and here even the great nineteenth-century ornithologist John Gould does not quite succeed.

OPPOSITE
There are some 340 species of hummingbird. All are small, feed on nectar and have incredibly fast wingbeats.

RIGHT
Only when a hummingbird is caught by an artist motionless can its iridescent colours be enjoyed fully, as captured in this nineteenth-century illustration of a Black-eared Fairy by British ornithologist William Swainson.

OPPOSITE
Hummingbirds feed on energy-rich nectar, here from a large sensitive plant, a form of mimosa, pollinating the flowers in the process. They can hover with great precision and can even fly backwards.

migratory, leaving their breeding grounds as plants cease to be in flower. Rufous Hummingbirds migrate to Mexico and pass along the Rocky Mountains on their 10,000-km (6,200-mile) return journey. Originally this migration would have been fuelled naturally by flowering plants, but many people now set up feeders in their gardens, providing sucrose solutions to mimic the plant nectar.

Some Rufous Hummingbirds are changing their migratory behaviour because of this. Increasingly they are migrating southeast in autumn (fall) to winter in southern US states such as Florida. One bird was ringed (or banded – a tiny metal, numbered ring put on the bird's leg) in January 2010 in Florida and recaptured in Alaska in June – a distance of 5,600 km (3,480 miles). Studies have also revealed that individual hummingbirds return to the same patches of flowers to feed in winter and summer, despite the enormous distances involved.

Reinagle pinxt. Stadler sculpt.

Large Flowering Sensitive Plant.

PEREGRINE FALCON

Falco peregrinus

ABOVE
These majestic birds have long been prized in falconry due to their ability to chase and catch fast-flying prey on the wing: the *Book of St Albans* in 1486 contained the phrase 'A Peregrine for a Prince'.

OPPOSITE
The Peregrine Falcon is the fastest bird: in dives at prey, speeds of 320 kph (200 mph) have been measured. But in between these short periods of intensity, Peregrines spend a lot of their time sitting, watching and waiting for vulnerable prey to appear.

The very name 'peregrine' derives from a word meaning 'wanderer' or 'travelling' and alludes to the fact that Arctic-breeding Peregrine Falcons migrate south in the winter, often to feed on large flocks of waterfowl wintering around estuaries. But most Peregrine Falcon populations, which are found on all ice-free continents, are resident rather than migratory, and it is the Peregrine's speed of flight, and the fact that it is a predator of birds on the wing, that mark it out as remarkable.

Outside the breeding season, Peregrines spend much of their time just sitting around, usually on a high perch. They may look rather lazy, but in fact they are ever-watchful, scanning the skies to spot likely prey such as pigeons or ducks flying in the distance. And once in flight, the Peregrine is transformed, taking on the appearance of great power and confidence. The broad-based wings speed the birds through the air with an impression of purpose and more than a hint of menace – they look like the assured and efficient killing-machines that they are. Often Peregrines will dive at a flock of birds and pull out at the last moment, as if to test the mettle of their potential prey; they may be assessing which of a flock appears to be the weakest bird and therefore the easiest victim. They prey on a huge range of birds, large and small.

The dive of a hunting Peregrine is called its 'stoop' and the bird falls vertically with wings closed, in a teardrop shape, at astonishing speeds of up to 320 km (almost 200 miles) per hour. The forces on the falling bird are immense. A Peregrine's nostrils contain baffles, rather like those of jet engines, so that the wind entering its airways does not explode its lungs. Around a third of Peregrine stoops result in a successful capture. Imagine the skill involved in

Louis Agassiz

OPPOSITE
Peregrines often nest on cliff ledges (increasingly, these days, on city buildings too). This illustration by the American ornithologist Louis Agassiz Fuertes shows the red-blotched egg, two young chicks, an immature young bird (the brown bird on the upper ledge) and an adult. Their prey here is a Meadowlark.

BELOW
The Peregrine Falcon was greatly affected by agricultural pesticides between the 1950s and 1970s. Being at the top of the food chain, Peregrines were particularly vulnerable and suffered from eggshell thinning, increased mortality and rapid population declines. Nowadays, with regulation of these chemicals, Peregrine populations are doing well across the world.

snatching living prey in a deadly grip at high velocity – the Falcon must judge speed, direction and moment of attack perfectly to ensure the prey does not escape and also not to injure itself in the collision.

Peregrine Falcons form pair bonds for life, and the female lays a clutch of around three or four eggs. After learning to fly, young Peregrines need to learn to hunt. Their parents teach them some of the skills required by dropping dead prey for the young birds to catch in the air, and there is a lot of aerial play between siblings and parents. After three to four months young birds disperse and must fend for themselves. Four out of five adult Peregrines survive from one year to the next, but only half of those fledged each year survive to their first birthday – and the main cause of death is starvation. Only the most adroit and adept survive.

Peregrines have been highly prized in falconry for over 3,000 years, as much as any other bird except perhaps the larger Gyrfalcon (*Falco rusticolus*). They are a bird fit for a king and are regarded by falconers and birdwatchers alike as daring as well as supremely skilful. Their means of capturing prey seem, to our eyes, to combine courage with dexterity, and fairness with a clean kill. We cannot help but picture them as dashing, noble fighter pilots – the 'Top Gun' of birds. But even the most powerful creatures are vulnerable, and Peregrines were listed as Endangered when agricultural chemicals both increased their mortality and reduced their productivity by causing a thinning of their egg-shells so that eggs broke in the nest. Since the banning of these chemicals Peregrines have made a comeback and are now a fairly familiar sight in many big cities including Brisbane, London and New York, where they may be found nesting on tall buildings, chimneys and structures that imitate the sheer cliffs they prefer in nature.

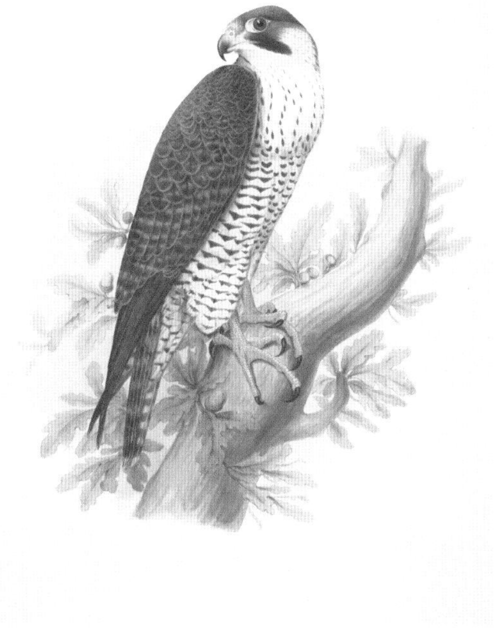

The Love Life of Birds

Mute Swan • Bullfinch • Phalaropes • Guianan Cock-of-the-Rock
Birds of Paradise • Dunnock • European Bee-eater
Brown-headed Cowbird • Ruff

ABOVE Ruff.

Many birds live their adult lives, whether long or short, with the same partner of the opposite sex, with whom they attempt to bring up their young. Birds face many of the same important questions when choosing a partner as do we – is he (or she) good looking? Sound good? Have lots of land? Likely to be a faithful partner? Will they be good with the kids? And things may not always run smoothly, so widowhood, divorce and infidelity are also features of the love life of birds, as they may be for us.

The Mute Swan and Bullfinch exemplify the basic model of a pair of birds mating for life and bringing up their young in a full and equal partnership. But there are numerous variations on, and departures from, this theme. The Brown-headed Cowbird of North America lays its eggs in the nests of other species and does not need to form permanent pair bonds as it lets others bring up the chicks. In some species, illustrated by Birds of Paradise in Papua New Guinea, the Cock-of-the-Rock from South America and the Ruff from Europe and Asia, the males play no part in the incubation of the eggs or the care of the young – all that is done by the females. The males gather at traditional display areas, known as 'leks' (from the Swedish for 'play'), where they compete, aggressively or showily, for the attention of the females. In almost all cases some males are very successful at mating with females, whereas others fail completely.

With Phalaropes, and just a few other species – only a tiny proportion of birds – the boot is somewhat on the other foot. The females are the more aggressive and brightly coloured gender and compete for the males' attention. It is the males who then incubate the eggs and bring up the young, while the females seek other partners and may lay several clutches of eggs.

A rather unassuming and drab-looking European woodland and garden bird, the Dunnock has perhaps one of the most varied domestic arrangements: from females with several males to males with several females and involving infidelity and even infanticide.

The European Bee-eater nests in colonies, and although most of the nest burrows are attended by a traditional pair of birds, some are helped by a third, fourth or even a fifth bird. This is a model which, with many variations, is fairly common among many tropical species.

MUTE SWAN

Cygnus olor

Mute Swans, with their pristine white plumage and orange bills, are long-lived aquatic birds that seem to epitomize avian love and faithfulness. As the members of a pair display, they move towards each other with their heads down, their two long, curved necks forming the shape of a heart – to the human eye at least.

Native to Europe and Asia, but now widespread elsewhere, Mute Swans (which are not in fact mute, they occasionally growl) are some of the largest flying birds in the world, and can live for over 20 years; a 10-year lifespan is perfectly normal. During this time Mute Swans usually remain with the same partner while both members of the pair are alive. To raise their family they build a large nest of vegetation on the ground, and both male (called a cob) and female (pen) will fiercely guard the nest-site from all predators and intruders, including humans, by hissing and raising their powerful wings.

Incubation of the clutch of eggs, which takes 40 days, is by the female (occasionally the male takes a turn), but when the young, called cygnets, hatch, they soon take to the water and are guarded and accompanied by both parents. The buff-coloured cygnets, up to nine of them, stay with their parents for many months, and generally first breed at four years of age.

Most Mute Swans have one mate in their lives, certainly for up to 14 years on occasion, and long-term studies confirm that they rarely choose a new mate if their existing partner is still alive. The death of the previous partner is usually the cause for finding a new mate, rather than divorce, but occasionally one member of a pair will cease breeding, perhaps from old age, and the other will seek a new partner. The divorce rate is about 4 per cent per annum (far lower than for humans in most countries).

Young pairs, and new couples (after the death of one of an existing pair), generally raise fewer cygnets than experienced pairs, so there does appear to be a real benefit in maintaining a lasting relationship. Mute Swans defend a

OPPOSITE
Swans are great examples of avian fidelity and form long-lasting pair bonds. Three species are illustrated here: the Mute Swan, adult and cygnet (foreground); Bewick's Swan (*Cygnus bewickii*, top left) and Whooper Swan (*Cygnus cygnus*, top right).

Pl. 45

Bewick's Swan. Whooper Swan.

Mute Swan (adult & young).

RIGHT
A symmetrical design by Victorian artist Walter Crane of two Mute Swans, with the curved necks and heads of the birds forming the shape of a love heart. This graceful pose is often replicated in nature, with pairs of Mute Swans facing one another and bowing in courtship.

stretch of lake or river from other pairs. It may be that established pairs, as well as being better parents, are also more skilled at repelling intruders from their territory efficiently – they form a better fighting team.

The story of *Swan Lake* is woven around the faithfulness of the swan. Odette has been transformed by a sorcerer into a swan by day. The spell can only be broken if one who has never loved before swears undying devotion to her. Siegfried, who is seeking a wife, is chasing a flock of swans in order to shoot one, but desists when the swan changes into the beautiful Odette at dusk. The two fall in love, but the course of true love does not run smooth as the sorcerer disguises his own daughter, Odile, as Odette at a ball. Thinking that Odile is Odette, Siegfried proclaims his love for her in public. Odette is resigned to dying of a broken heart, but Siegfried realizes he has been tricked and chooses to die with Odette rather than live with Odile. The faithful and love-smitten couple throw themselves into the lake and ascend to heaven together.

The Mute Swan is an avian version of an idealized human way of living. They have their own plot of territory, mostly water, and make a home where they rear their young. The male and female are very much a partnership, looking after their family together and co-operating, although having slightly different roles in the pair. And they are faithful to each other through life.

BULLFINCH

Pyrrhula pyrrhula

If there is an ideal model of mating behaviour in birds then the Bullfinch of Europe and Asia might be considered to come close to it. Throughout the year, males and females are almost always seen together, suggesting that the pair bond remains strong. Recent studies of the internal morphology of male Bullfinches have shown that they have, for their overall body size, relatively small testes, which is taken as support of the idea that this species is, indeed, highly faithful to its mate.

Male and female Bullfinches are similar in size, although the male is a little larger and has a brighter plumage, with a distinctive pinkish-red breast and grey and black upper parts. The female could be said to look like a slightly faded version. Both parents share the task of feeding insects to the young in the nest and a pair can raise two or three broods in a season, with generally four to five eggs laid each time.

The Bullfinch, whose name comes from its thick neck and blunt head (apparently resembling a bull), appears a somewhat shy bird, with a very quiet song that is rarely heard, and a sad and rather plaintive call. For most of the time it stays close to dense vegetation and, unlike other finches, rarely visits gardens to use bird-feeders. It mainly feeds on seeds and berries, but in spring its short, broad beak makes easy work of blossom buds on fruit trees. In most parts of its range the Bullfinch is a resident species and does not migrate, adding to its image as a settled and stay-at-home type of bird.

ABOVE
Bullfinches are almost always seen in pairs or very small, perhaps family, parties. This illustration appears to show a male (above) and a slightly duller female (below).

BELOW
The four or five eggs in a clutch laid by a female Bullfinch are likely to have been fathered by the male who accompanies her. It is thought that extra-pair fertilization is very low in this species.

PHALAROPES

Phalaropus tricolor, Phalaropus fulicarius & Phalaropus lobatus

A cursory glance at any of the three species of phalarope in the world – Wilson's, Grey (or, confusingly, Red) and Red-necked – would probably not immediately indicate anything very special about this group of elegant wading birds, with stilt-like legs and fine, long, pointed bills. It might be assumed that their most unusual behaviour is that they can swim on the water surface and spin round, creating a miniature whirl of water around their bodies, which brings small food items, such as floating insects, closer to their reach. It's unusual, and fun to watch, but not that outstanding. Or perhaps it is that in the non-breeding season they migrate south and spend most of their time (Grey and Red-necked Phalaropes at least) at sea, unlike most wading birds.

However, closer study would reveal that in all three phalarope species the sexual roles and breeding system are very different from those of most other birds. It is the female phalaropes that are the more colourful and slightly larger gender, and they display and compete among themselves for nesting sites and the attention of males. Females also call more than males and bicker among themselves, and it is the females who fly over the marshes in display flights. After mating, the males carry out all the incubation of the eggs and parental care of the chicks, with no help from the females at all. Some females may seek multiple mates during the short Arctic breeding season, but play no part in rearing any of their young after egg-laying.

Whereas there are plenty of species where the males have little or no involvement in parental care, the opposite situation is very unusual. The three phalaropes, together with the Dotterel and Spotted Sandpiper (both also wading birds) and a few species of jacana (tropical waders), are the main species to exhibit this behaviour.

In lekking species such as the Ruff, Cock-of-the-Rock or birds of paradise, the liaison between the male and female lasts only the length of time it takes to mate, but with phalaropes the pair remain together for about seven to ten

OPPOSITE
Two of the three phalarope species are seen here, each showing the breeding (bright) and winter (generally grey) plumage. The birds at the top are Red-necked Phalaropes, and the ones below, Grey Phalaropes. Both species live in northern, sometimes Arctic, wetlands, but winter at sea far to the south.

♂ 4/5 ♂

♀ 1/2 3/4 ♀

ABOVE
Wilson's Phalarope is named after the eighteenth-century Scottish bird artist Alexander Wilson, who produced one of the first books of American birds. With this North American wetland bird, as with the other two phalarope species, the females (right) have much brighter plumage than the males (left) and take initiative in display and competing for mates.

days. During this period the pair court, mate and excavate a simple scrape of a nest cup in the ground; the clutch of four eggs is laid at a rate of no more than one egg each day. While in this arrangement some sort of a bond does develop between the two birds, at the end of laying the female leaves her more cryptically plumaged mate (who is around 25 per cent smaller) to get on with incubation and child-rearing and she moves on, sometimes to form another bond with a new male. Only a small proportion of phalarope females, around 10 per cent, actually pair again with second males, so although the roles of the sexes are reversed, this does not always lead to females having multiple mates. If a female does secure another mate then she may lay her second clutch about a week after she completed her first.

Quite why this role reversal occurs in just a few species is not yet fully understood. It is indeed rare, and closely related species such as the Spotted Sandpiper of North America, in which sex roles are reversed, and the almost identical Common Sandpiper of Eurasia, in which they are not, display different patterns. Sex role reversal seems commoner in wading birds, but it crops up at low frequency in other groups of birds with quite different ecologies. It remains a puzzle.

GUIANAN COCK-OF-THE-ROCK

Rupicola rupicola

ABOVE
A Guianan Cock-of-the-Rock, depicted accurately as far as plumage is concerned, although deep rainforest is the natural habitat of this bird. Males display with their brightly coloured plumage and compete for the attention of females.

Living deep in the rainforests of French Guiana, Suriname, Guyana, Venezuela, Colombia and northern Brazil, at altitudes of 300–2,000 m (900–6,000 ft), male Guianan Cocks-of-the-Rock are among the most brilliantly coloured of all birds. Their blazing orange plumage, with wispy feathers appearing from the wing and tail, together with a strange disc-shaped crest, stands out startlingly against the green leaves among which they perch and display. The females by comparison are somewhat drab, though both sexes have a round yellowish eye that can give them a rather piercing look. Both sexes have a catholic diet of fruit, but small reptiles, amphibians and insects are also taken.

Males gather together at communal display grounds and show off their gorgeous plumage to females, who visit these arenas to make the choice of who will father their offspring. The females are then solely responsible for all parental care, and their plumage is subdued because they need to avoid predators while incubating the two eggs in the nest they build of mud, vegetation and their own saliva in caves or on rocks. The same nests are used from year to year.

The males never see a nest, however, nor the eggs or young that they fathered. Their breeding season is totally dominated by displaying to attract females to mate, and competing among themselves to establish their dominance, or subordinance. This system has not been studied in great detail, but the expectation is that at each gathering of males, one or two of them will mate with most of the females that visit. Since Guianan Cocks-of-the-Rock are long-lived, a male may have several years contesting to be the successful male – presumably some never make the grade, while others are spectacularly successful.

It is as though all the male's energies are channelled into looking good. The males flick their wings open and crouch low over the branches of the trees. Their occasional calls are rather like those of a squealing pig. Males also interact with each other at times, appearing to confront each other, flicking their wings and snapping their bills, and calling loudly. But, at least usually, this is all just for show – perhaps with the serious intent of establishing a dominance hierarchy. Actual violence seems to be threatened but rarely, if ever, carried out.

The same gathering places are used repeatedly from year to year, and since the gaudy plumage of the males is the antithesis of camouflage, there can be no doubt that these birds want to be noticed. It might be conjectured that the competing males engaged in their energetic displays would be more susceptible to predators, and this may be one way that a subordinate male can rise up the pecking order. Even during the middle of the day, when display is at a much lower level of activity, the glowing colours of the male birds must surely make them more visible predators such as eagles, falcons, jaguars and boa constrictors, as the birds search the forest for fruits on which to feed. Perhaps while the brightest coloured males are most visible and have greatest access to females, they also suffer more losses to predators – and so the balance of those two factors could determine the gaudiness of the male Guianan Cocks-of-the-Rock, but constrains their plumage from being even more extreme. What we may be seeing here is an evolutionary compromise.

In the rainforests of northeast South America, the dazzling male Guianan Cocks-of-the-Rock strut to show off their plumage, while the females go quietly about their business of bringing up a family. It's an uneven share of responsibilities, but this division of labour is a success for this species and for both sexes; they have been living in this way for thousands of years.

BELOW
A female Guianan Cock-of-the-Rock has much browner plumage than the splendid males. The females incubate the eggs and raise the young birds; the males display to attract more mates and take no part in parental care.

OPPOSITE
The Guianan Cock-of-the-Rock is a gorgeous specimen when in full display. As well as the bright orange plumage, its half-moon crest and wispy feathers create one of the most striking avian sights.

J. Gould & W. Hart del. et lith.

PARADISEA PAPUANA.

BIRDS OF PARADISE

Paradisaeidae

The birds of paradise comprise about 40 species of crow-sized birds, most of which are found in Papua New Guinea. Indeed, they are so associated with this country that a Raggiana Bird of Paradise appears in flight on the nation's flag in yellow outline, showing the extravagant flank plumes of the male. As in all birds of paradise, the male is very striking, while the plumage of the female is duller, as she needs to be camouflaged when incubating the eggs and caring for the young. The males have no such constraint, as their role is limited to copulation with females and they play no part in parental care. They have therefore developed a spectacular range of feathers in dazzling shapes and colours, as well as extraordinary mating displays.

Darwin puzzled over how such elaborate and colourful plumages could evolve through natural selection and decided that their existence

> *... depends, not on a struggle for existence, but on a struggle between the males for possession of the females; the result is not death to the unsuccessful competitor, but few or no offspring.*

And that remains the scientific consensus today. Perhaps the plumage is actually a signal to females that the male is fit, sexually mature and healthy.

The feathers of birds of paradise have long been highly prized and used as adornments and in rituals in their native lands. Prepared skins, lacking wings and especially feet, were often traded, and some found their way to Europe in the sixteenth century. These beautiful, unfamiliar, and apparently wingless and footless birds, aroused great interest and scientific speculation. It was thought perhaps that they floated high in the sky, maybe in Paradise itself, where they

ABOVE
The Magnificent Bird of Paradise is indeed truly magnificent – at least the males are (top left and below). The females choose their mates on the basis of their display and plumage, and then incubate the eggs and feed the young. They have no need to look so spectacular, and every reason to be cryptic in plumage and behaviour.

OPPOSITE
Each male Lesser Bird of Paradise is a frenzy of colour and plumage as he displays to a watching female. Four males are depicted here by John Gould, while a plainer female visits the communal display to make her choice of mate.

RIGHT
The first birds of paradise seen in Europe in the sixteenth century were in the form of prepared skins without feet or wings, in order to display their plumes to best effect. Scholars speculated that these birds floated in the air and never came to land. This illustration of skins, one with feet, is from Francis Willughby's *Ornithology* of 1678.

OPPOSITE
The Greater Bird of Paradise, here depicted by a nineteenth-century Japanese artist, is the largest of the 40 bird of paradise species.

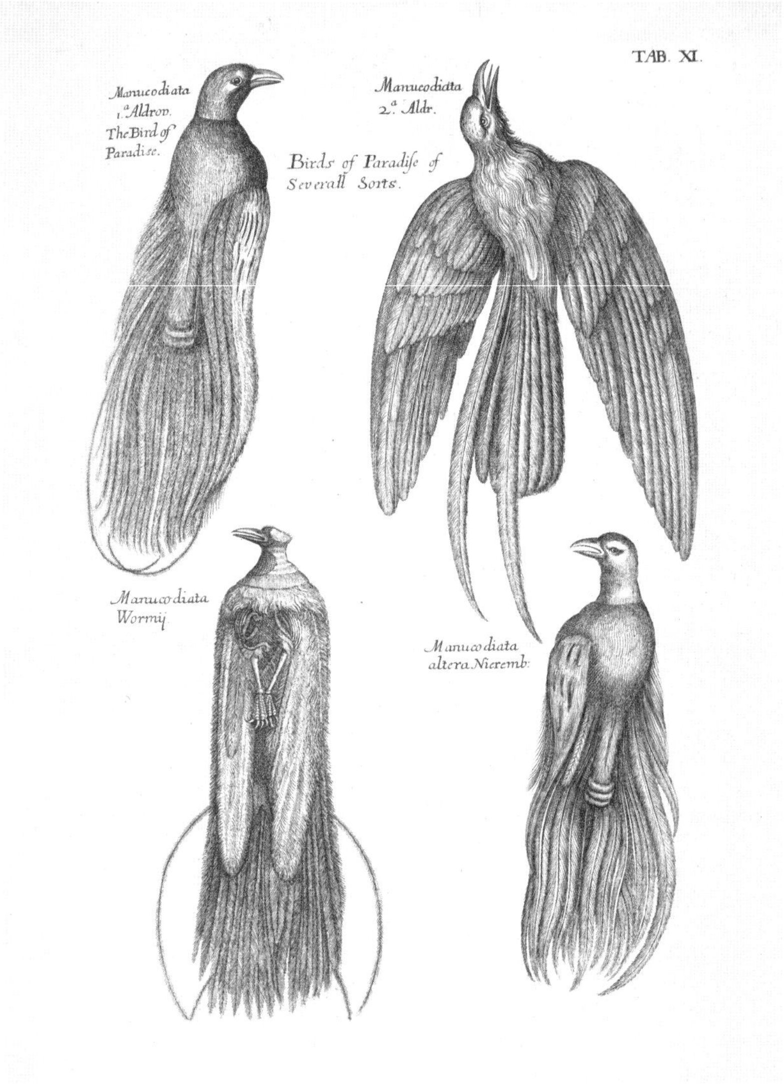

fed in the clouds. The great Swedish taxonomist Carl Linnaeus named the Greater Bird of Paradise *Paradisaea apoda* – *apoda* meaning 'without feet'.

Residents of Papua New Guinea knew, of course, that these birds displayed at traditional sites on the forest floor or in trees, where males compete and display their plumage to females who visit to choose mates. The displaying males with their contortions, dance-like movements and posturing are simply beautiful and astounding. Although it is the male birds of paradise we admire, this is the product of female choice rather than male creativity, as female birds of paradise have selected more and more extreme varieties of male.

フウテウ

DUNNOCK

Prunella modularis

Appearances can be deceptive in the bird world as elsewhere. The Dunnock is a mousy little bird – it creeps around in the undergrowth pecking away at small insects, and with its brown and grey plumage it doesn't do much to draw attention to itself. But its domestic arrangements are the stuff of soap operas.

Female Dunnocks are a little unusual in that they defend their own feeding territories in the nesting season. Every bit of woodland or garden occupied by Dunnocks is divided neatly into female plots of land, where they will feed and nest. Male Dunnocks defend territories too, but they aren't the same as those of the females; the males divide up the nesting habitat in their own system and so the sexes have their separate territories, which overlap in a range of different ways. Sometimes two males will defend a joint territory, with one male being dominant and the other rather subordinate.

Not surprisingly, males mate with females who live in their territories, and females mate with males who live in their territories, but because the territories aren't the same this can lead to a variety of pairings. About 40 per cent of Dunnock breeding arrangements involve one male and one female – much like the vast majority of bird species. But the next commonest arrangement is for two males to share a single female, and that makes up about 30 per cent of breeding groups. The third commonest arrangement is for two males to breed with two females (where a shared male territory overlaps with the territories of two females), and that accounts for a further 17 per cent of breeding groups (not very common, but not very unusual either). Aside from these three most frequent combinations of males and females there are many others that scientists have observed, with one to three males paired with one to three females in all sorts of combinations (and probably others that haven't yet been observed by voyeuristic scientists).

OPPOSITE
The rather mousy Dunnock is a small grey and brown bird that hops around in garden undergrowth, not drawing much attention to itself. However, it has a mating system that is more complex than first meets the eye, with a variety of breeding arrangements.

Male Dunnocks are good fathers – provided they 'think' they are the fathers of the offspring they are feeding. When the chicks hatch, males only

Stab. cromolit. O. Dressler Milano

ABOVE
Dunnocks have one of the most complicated mating systems of all birds. Males and females each defend territories, sometimes with another member of their sex, and breeding groups can comprise a small number of males (1–3 most often) with a small number of females (1–3 most often), but in many different combinations.

feed the young of females with whom they have mated, and in the case of pairs of dominant and subordinate males sharing a female, the dominant male will have fathered 55 per cent of the chicks in the nest and the subordinate male just 45 per cent of them. The males adjust their feeding rates at nests in relation to the number of times that they mated with the female. In groups of two males with two females, each male helps feed the young of just one female even though both males would have mated with both females.

Even this account of the private life of the Dunnock only scratches at the surface of this system, and years of study were necessary for scientists to uncover its complexities. But there is yet another aspect of the Dunnock's home life, which requires rather sensitive description. After mating, a female bird stores the male's sperm for several days and releases it gradually to fertilize her eggs before they acquire a hard shell. Male Dunnocks stimulate the females to eject stored sperm before mating with them in order to increase their own chances of fertilizing all future eggs. This stimulation takes the form of pecking at the female's rear before mating, and is unlike anything known for any other bird species.

EUROPEAN BEE-EATER

Merops apiaster

True to its name, the European Bee-eater does indeed feed on bees, and other large insects, which it catches in flight. A single bird can consume over 200 insects in a day, making them not very popular with beekeepers. In April and May when the birds return in large numbers to Europe and Asia from Africa, where they have spent the northern winter, the colonies are alive with these brightly coloured, exotic-looking birds. They nest in colonies, with pairs excavating deep nest burrows in banks and cliffs.

ABOVE
One of Europe's most colourful breeding birds, the European Bee-eater returns in flocks from its African wintering grounds in late April and May. The arrival of these birds, with their piping flight calls, is a sign that spring is giving way to summer.

In the early stages of the breeding season the birds spend a lot of time perched in pairs on prominent features such as telegraph wires. Careful observation will show that the females, whose plumage is just slightly greener than that of the males, sit still while their mates fly out to catch passing insects, which they bring back to the pair's perch. Large insects, such as dragonflies, are bashed on the perch to kill them and then fed to the female. Smaller prey items, such as bees, are also stunned or killed at the perch, and in the case of bees the sting is removed by a clever wipe of the bee's abdomen on the perch before being swallowed, usually by the male himself. This is the period of egg formation, and the males provide the resting females with food so that eggs can be produced and laid as quickly as possible. Perhaps during the period when they are forming eggs the females are also less manoeuvrable in flight: to catch a fast-flying bee or dragonfly on the wing requires speed and dexterity.

Incubation of the clutch starts when the first egg is laid, with the females spending the night on the nest and the two parents sharing shifts during the day, the other feeding in or at a little distance from the colony. Because the eggs are laid about a day apart, and the clutch may have up to nine eggs, the last chick hatches when its oldest sibling is nine days old. This form of incubation

ABOVE
European Bee-eaters nest in colonies, with each pair digging a nesting tunnel that leads to a chamber in which up to nine eggs are laid. When the young are being fed, their parents are sometimes joined by other birds, called 'helpers', who assist in provisioning the chicks.

helps reduce the 'peak need' of food for the chicks and may be an adaptation to unpredictable food supplies.

With the adults rushing in and out of each nest with insects to feed their young, the colony is a very busy place at this time of year. But then at some nests, extra birds start to provision the chicks with food. No longer is it just the male and female parents at the nest – so-called helpers arrive and share the work, behaviour that was noted by Aristotle in his *History of Animals*.

Only around one in twenty bird species have 'helpers at the nest' and many of these occur in Australasia and southern Africa, but the phenomenon crops up in a wide range of other species across the world. The helpers are usually closely related to the pairs that they help – it's a phenomenon that evolves not in error, and not out of altruism, but through natural selection favouring those birds who help their relatives (who share their genes) to produce offspring. Often such helpers are birds that are too young to breed or can't find a nesting territory of their own, so 'stay at home' with their parents and make themselves useful. But that's not the case with European Bee-eaters, since there are plenty of sites for additional nest burrows in their colonies.

Almost all the helpers at European Bee-eater nests are male birds, and studies have shown that they are individuals whose own nesting attempts have

RIGHT
European Bee-eaters do indeed eat a lot of bees, which they catch on the wing and take to a perch before removing the sting by dexterously wiping the bee against a branch. Other insects such as beetles and dragonflies are also eaten, but bees predominate.

failed, sometimes through their nests being predated by snakes; if it's too late to create new nests then birds will help at those of relatives in the colony. But why is it usually males – why don't females so often help their relatives? The answer lies in knowing that in this species, as in many birds, the males return to nest close to where they were hatched and raised, usually in the same colony, whereas females disperse to new colonies. Therefore it is most often males who can find relatives nesting near them if their own nest attempt fails. Male European Bee-eater helpers are usually assisting their own parents or their brothers to raise their young; the occasional female helper is a mother helping at the nest of her son from a previous year.

BROWN-HEADED COWBIRD

Molothrus ater

ABOVE
The Brown-headed Cowbird is a seed-eating bird of North American grasslands, although thanks to felling of the native forests it now inhabits farmland in previously forested areas. There is no permanent pair bond between the sexes, and the females are obligate brood parasites, laying their eggs in the nests of a wide range of other species.

A small proportion of bird species (about 5 per cent) are what is known as obligate brood parasites – put simply, they lay their eggs in other species' nests where the 'hosts' then rear their young. Familiar examples include some New World and Old World cuckoo species, but perhaps less well known is the Brown-headed Cowbird of North America.

Brown-headed Cowbirds are primarily grassland species and avoid forests, and so would once have been largely restricted to the Great Plains of North America, where they would have fed on seeds and on insects stirred up by enormous herds of bison. With the felling of the native hardwood forest east of the Mississippi River and the cultivation of land, Brown-headed Cowbirds could spread much further east and interacted much more frequently with forest-dwelling birds, including the Wood Thrush (p. 23).

Female Brown-headed Cowbirds are now known to lay up to three dozen eggs in the nests of other species in a single season, one in each nest. All the females' efforts are put into finding suitable nests, and then laying as many eggs as possible. Overall, Cowbird eggs have been found in the nests of over 200 different host species, but it is now thought that individual Brown-headed Cowbirds specialize, at least to some extent, in particular favoured host species, which might be the much smaller Yellow Warbler or the slightly larger Red-winged Blackbird.

This unusual behaviour has the advantage that the cowbirds are spared the effort of first building a nest and incubating eggs and then tending their young, and this is a reproductive system where both parents have no real contact with their offspring. Unsurprisingly, therefore, cowbirds do not form permanent pair bonds, and both males and females mate with several partners.

The female Brown-headed Cowbird does, though, have to find a large number of host nests in which to lay her eggs, one at a time. Once she has located a suitable nest, she removes an egg as she lays her own, to reduce the chance that hers is spotted as an intruder. The cowbird egg hatches very quickly, allowing it to get a head start on the host's actual offspring when the parents bring food to the nest. Sometimes the young Brown-headed Cowbird will eject unhatched host eggs from the nest, and even other chicks, in order to get a greater share of the food brought by the devoted parents.

It's hard not to think in human terms and see the cowbirds as behaving unfairly and taking advantage of their innocent hosts. But the truth is that this behaviour has evolved, rather uncommonly, because it is a successful way of making a living. It's also difficult not to regard the hosts as being a bit stupid for not recognizing that the Brown-headed Cowbird nestling is not of their own species. However, a bird species that carelessly tossed out eggs or chicks from its nest might throw out quite a few of its own babies along with the Brown-headed Cowbird's and so be worse off as a result. But the Cowbird doesn't have it all its own way, and certain host species are better at spotting what is going on than others, and respond in different ways.

RIGHT
A female Brown-headed Cowbird has laid an egg in the nest of another species as a male looks on. When the host returns, it may accept or reject the foreign egg. Some species have evolved better discrimination of Cowbird eggs than others, and an evolutionary arms race is played out between Brown-headed Cowbirds and a range of hosts across North America.

Some species, for instance the Blue-gray Gnatcatcher, simply desert their nests if they are parasitized, and so in theory the cowbirds should evolve to avoid those species. The Yellow Warbler is quite adept at recognizing a cowbird egg, but is too small to eject it, so instead buries it at the bottom of the nest under nesting material and it is not incubated. Other species, such as the Brown Thrasher, simply puncture the cowbird egg as well as often ejecting it from the nest. So it is as though there is an ongoing arms race between Brown-headed Cowbirds and their hosts. If all hosts evolved skills of efficient identification and ejection of cowbird eggs, then the Brown-headed Cowbird would be in danger of extinction – but it hasn't happened yet.

Brown-headed Cowbirds exhibit two more types of apparently crafty behaviour that can be seen as part of this arms race. One is 'farming', whereby they destroy host nests that are too well advanced to parasitize in order to force the host to lay more eggs and so give the cowbird a chance of parasitizing a future nest. And secondly they exhibit 'mafia' behaviour. This is when they revisit nests where they have laid eggs and destroy them if their eggs have been ejected by the host – a form of avian retaliation, penalizing those hosts who show discrimination against the Cowbird egg.

From a human point of view we disapprove of the behaviour of the Brown-headed Cowbird, as it seems to us to be manipulative and exploitative. Similarly, we sympathize with the hosts and raise a cheer for those who 'fight back' and try to turn the tables on the brood parasite. But if we leave inappropriate moral judgments aside, then it is impossible not to marvel at the complexity of the evolutionary arms race that is being fought out in the nests of the hosts of the Brown-headed Cowbird.

RUFF

Philomachus pugnax

ABOVE
The feathers that give the Ruff its name recall the dress of Elizabethan courtiers. Only the males have this extravagant plumage, and only in the breeding season. The smaller female is known as a Reeve, as well as a female Ruff.

It is fairly common for the males of bird species to be larger than the females, but the size disparity between the sexes of this migratory wading species is so great that it has led to them having individual English names: Ruff and Reeve. The male Ruff is about 75 per cent heavier than the Reeve and gets its name from its extravagant breeding plumage: the males grow distinctive tufts of feathers on their heads and large thick groups of feathers on the breast that resemble the exaggerated ruffs of sixteenth-century courtiers.

Males gather to display at their leks in spring, where they defend very small bare patches of ground in grassy areas. The Latin name for the species is *pugnax*, meaning 'combative' or 'pugnacious', and the male Ruffs certainly live up to that. Up to 20 males are commonly found at any particular lek, and only the oldest and strongest can hold on to central, prized display areas. From day to day, the same males return to the same display stands, with a certain amount of testing of strength each time.

When the Reeves visit the lek, the males stop their fights and freeze into a crouching display. Reeves inspect the static males and then crouch to invite a chosen male to mate with them, the central males gaining most mating opportunities. At leks where the male dominance hierarchy is not well established and fighting breaks out, females tend to be deterred from mating. It is as though the Reeves expect the males to have already sorted out their ranking.

The system thus far described is fairly typical of lek species – brightly coloured or adorned and aggressive males display to impress females, who then choose to mate with the dominant males. But with the Ruff there is a further twist to the story. Another group of males have a different strategy. These are called satellite males and they have similar plumage to other males, but their ruffs and some of their body feathers tend to be white. They are therefore identifiably different. About one in six male Ruffs are 'satellite males'. Such satellite males have been noted in many parts of the world where Ruff leks have been studied, so they appear to be a constant part of the breeding system.

Satellite males also behave differently – they do not compete for females by fighting and rarely display aggressive behaviour. They simply attempt to mate with females that visit the dominant resident males, and sometimes succeed. It seems that the dominant males tolerate the satellite males because the females prefer leks with lots of males, thus increasing the number of visiting females.

But it gets even more complicated – there is a third type of male, called a faeder male. These closely mimic Reeves in plumage but also show some male characteristics, though not the extravagant ruffs. They are intermediate in size too. These males attempt to escape the notice of the resident males and then to 'steal' mating opportunities with Reeves visiting the dominant males.

The sexual dimorphism of the Ruff (and Reeves) is impressive enough, as are the antics employed by resident males, but we now know that there are two other strategies employed by some male Ruffs as alternative routes to mating success, making this bird's love life one of the most intriguing.

ABOVE
Male Ruffs display to females at traditional display sites known as leks. The males compete for the best spots, and the females visit these arenas simply to choose a mate.

OPPOSITE
Male Ruffs vary considerably in the colour of their head tufts, wings, backs and ruffs, meaning that they can be recognized as individuals even by the human observer.

OVERLEAF
Ruffs at a lek display to a visiting female, or Reeve. Note the bird with the white ruff, which probably signals that he is a 'satellite male'. The males stand motionless as the visiting Reeve makes her choice of mate.

4
3
1

Avian Cities

Flamingos • Pelicans • Lapwing • Red-billed Quelea
Rook • Atlantic Puffin • Emperor Penguin

ABOVE Flamingo.

There are advantages and disadvantages to living in close proximity to large numbers of your fellow species, as any human city-dweller might agree. Most birds are territorial, and may defend their feeding and nesting space fiercely, but about one in ten bird species is colonial.

Birds choose to nest together for two main reasons. The first is when food supplies are concentrated in small areas (this is the case for wetland species including flamingos and pelicans, as well as the grassland-living Red-billed Quelea), and the second is because safe, predator-free nest-sites are few and far between (which is why Atlantic Puffins nest on predator-free islands, but it also applies to flamingos and pelicans again).

Clearly, nesting with thousands of others increases competition for food, so it is only where food supplies are superabundant, for instance some wetlands, that coloniality can be a suitable strategy. But the mere presence of a multitude of others also reduces any one individual's risk of falling victim to a predator. This is certainly true in some of the largest bird colonies on Earth, those of the Red-billed Quelea, which can number millions of nests. Predators may have a bonanza of eating eggs, chicks and adult quelea, but the overall proportion that is taken is very small.

Some species benefit from colonial living in other ways. Pelicans often feed communally, and together a flock of pelicans can co-operate to drive and catch fish that a single bird would never be able to. Emperor Penguins huddle together for warmth in the freezing Antarctic winter as they incubate their eggs, and later their growing chicks also save energy by huddling together. Lapwings will attack any predators of their nests or chicks, and a larger colony of birds is more effective at this than a lone pair or small group – there really is strength, as well as safety, in numbers.

Reasons why other species, such as the Rook, nest in colonies are a little more difficult to fathom. Rooks feed sociably throughout the year and may find some advantage in that, and so it would seem natural to nest together too. Whatever the reasons, where large numbers of birds gather together to nest the sight, sound, and sometimes the smell, can provide an impressive spectacle of magnitude, colour and movement, whether it's the sheer mass of a million quelea nests or the glowing pink of thousands of flamingos.

FLAMINGOS

Phoenicopteridae

All six species of flamingo, four in the New World and two in the Old World, are a shocking pink or red colour and all nest in large colonies. They are striking birds, with their very long legs and necks and curiously shaped bills, and the combination of bright plumage and coloniality make the sight of nesting flamingos one of the avian wonders of the world.

Flamingos are tall wading birds, with the largest species, the Greater Flamingo (*Phoenicopterus roseus*), measuring up to 150 cm (60 in.) in height. When walking, but perhaps even more in flight, they appear to be all neck and legs – with a relatively small body and short wings. This is a group of birds that is instantly recognizable from their colour and body structure.

The very name 'flamingo' comes from the Latin *flamma*, meaning 'flame' or 'fire', reflecting the colour of their plumage. Young birds in fact start life as a dull grey, and in a flock of adult flamingos it is possible to notice a great variation in the depth of hue, but en masse flamingos assault the eye with the brightness of their plumage. The colour derives from their diet of algae and shrimps, which contain carotenoid pigments. Captive flamingos are fed a diet including shrimp to ensure that they take on the characteristic pink colour.

The algae and shrimps are filtered from the water through the birds' bills, which are marvellously adapted to this method. Unlike in other birds, the lower bill is bigger than the upper, and the upper bill is not fixed solidly to the skull, so it is the jaw that can move. When feeding, the flamingo reaches down into the water with its very long neck, holding the bill upside down and moving it from side to side so that water passes into it. The fat tongue of the flamingo (which was regarded as a delicacy by the Romans) helps to pump the water and its contents through the beak. Numerous rows of horny plates inside then act much like the baleen plates of whales and filter the water so that small shrimps up to a couple of centimetres in length, but also microscopic algae, are extracted. Flamingos also use their feet to disturb the bottom sediment

OPPOSITE
Greater Flamingos, like other flamingo species, nest in huge colonies with the nests packed closely together.

and enrich the water they filter with more of the microscopic organisms.

Flamingos often form their colonies in alkali lakes and salt pans where the very high salinity encourages shrimps and blue-green algae to thrive in great densities, in turn allowing such large numbers of birds to congregate together and feed. The caustic nature of the water, to which the flamingos have adapted in various ways, also protects them from predators. Lake Natron in Tanzania is the only regular nesting site for the world's 2 million Lesser Flamingos (*Phoeniconaias minor*). They build a mound of mud, about 45 cm (18 in.) in height, as a nest, and this protects the single egg from small changes in water level.

Greater and Lesser Flamingos often nest in seasonal wetlands where the rains are unpredictable, and it has long been noticed that they arrive at sites such as Etosha in Namibia within hours of sufficient rain falling to create the right conditions for nesting. Hundreds of kilometres away, on the coast, where it is not raining, how do the flamingos know it is time to leave and fly such great distances? Perhaps they hear far-off thunder or read the clouds with more skill than we ourselves could. At present, we do not have a certain answer to the question.

All flamingo species display in tight groups of dozens or even hundreds of birds, walking together in the water with their long necks fully extended. There is head-shaking and some wing-flapping, and through these displays pairs are formed. Differences in the height and colour between individuals are very noticeable to the human observer, and presumably to the flamingos too.

From the Andes of Peru to the low-lying land of Egypt, flamingos have been regarded as sacred birds. We have admired them for their unusual morphology, their eccentric displays and mode of feeding, and for their elegance, and they are always a favourite in zoos and parks around the world. When they gather together in nature to nest in colonies, their swirling masses of pink and orange form an unrivalled avian spectacle.

ABOVE
This sixteenth-century illustration by John White shows that the flamingo was a familiar species even then, from birds kept in captivity at court and by wealthy landowners.

OPPOSITE
The very long legs and necks of flamingos make them immediately identifiable, even by those who have never seen the live bird in the wild or in a zoo. Flamingos acquire their pinkish colour from chemicals in their diet of algae and shrimps.

PELICANS

Pelecanidae

ABOVE
Pelicans were once thought to feed their young with blood by piercing their own breasts. These birds were often used in Christian iconography and became associated with the resurrection of Christ.

OPPOSITE
Pelicans, such as this Brown Pelican, nest in large colonies, often with other waterbirds. They feed socially, with a line of pelicans working together to herd fish towards the shore while dipping their beaks into the water in a co-ordinated manner to form a living fishing net.

Pelicans, with their long beaks characterized by a large expandable pouch hanging from the lower bill, are another very familiar-looking member of the bird world. The eight species of pelican nest on every continent except Antarctica and are all aquatic feeders, eating large fish and a variety of smaller prey; they all also nest colonially.

Nesting colonies can number thousands of pairs and are usually situated on predator-free islands, often shared with other aquatic birds such as geese, gulls and cormorants. The nest-sites are built either on the ground or in trees, depending on the species. The males defend them and attract females to nest with them. The two parents then share the incubation duties and both feed the chicks.

As well as nesting colonially, pelicans are unusual among birds in often feeding communally. Groups of pelicans will swim together, like a line of ships in tight formation, and beat their wings on the water to drive shoals of fish into shallows near the shore, where they can more easily be scooped into the pelicans' capacious bill pouches. Presumably, individual birds would be easily evaded by the fish and so the benefit of increased food availability more than outweighs the disadvantage of sharing the resource with lots of other pelicans.

The pelican essentially uses its expandable pouch as a net. Although single large fish can be grabbed and then manipulated and swallowed whole, another fishing technique involves scooping up as many small fish as possible at the same time in large gulps of water. Once caught, the water – as much as 10 litres (over 2 gallons) – is strained from the pouch until the fish can be swallowed. This process takes time, and other birds such as gulls, and sometimes even other pelicans, take advantage of this to steal the captured fish from the pelican's bill.

RIGHT
Pelicans catch small fish by scooping them up, en masse, into their capacious bill pouches and then straining out the water. The pouch can contain 10 litres (2 gallons) of water and the fish that were swimming in it. This drawing is by an early nineteenth-century Indian artist.

OPPOSITE
Most species of pelican usually feed while swimming on the water surface, sometimes upending like ducks, and dipping their beaks and heads into the water. However, Brown Pelicans and Peruvian Pelicans often dive into the sea from heights of up to 60 ft (20 m).

Pelicans walk without much style on land, swim effortlessly in the water but are perhaps shown to best effect in flight. They fly on broad outstretched wings – up to around 3 m (around 11 ft) in span – with occasional flaps, but often gliding on coastal updrafts or thermals, frequently in large groups in line astern. Their feeding grounds may be great distances from the nesting colonies.

North American pelicans, both Brown Pelican (*Pelecanus occidentalis*) and White Pelican (*P. erythrorhyncus*), were greatly affected by poisonous chemicals used in agriculture in the 1950s, particularly endrin, dieldrin and DDT. Both species have recovered very successfully since the use of these chemicals was banned and conservation efforts were put in place. Such a population recovery recalls the role of the pelican in early Christian symbolism. Due to a fanciful understanding of the pelicans' lifestyle including feeding their young with their blood by piercing their own breasts they became associated with the resurrection of Christ and are often depicted in churches and cathedrals.

Brown Pelicans in the USA and European White Pelicans (*P. onocrotalus*) can become very tame and often loiter around harbours and docks, where they can be seen at close quarters. They are also a familiar sight in central London's St James's Park, almost in the shadow of Buckingham Palace; pelicans have lived here since 1664, when the first arrived as a gift from the Russian ambassador to King Charles II. To stand close to one of the largest birds on Earth, and look into its eye, is a thrilling experience. However, in the wild they must nest in areas where there are relatively few predators and where they are within easy commuting distance of rich feeding grounds.

LAPWING

Vanellus vanellus

ABOVE
Lapwings nest on the ground in loose colonies. When they have eggs or chicks, all the members of the colony help to protect the young by 'mobbing' predators such as foxes or crows, launching repeated attacks from the air to drive them away.

BELOW
Lapwing eggs, or plover eggs, are regarded as a delicacy and were once easy pickings for hungry farm workers. The normal clutch comprises four eggs, laid in a simple scrape on bare soil or grass.

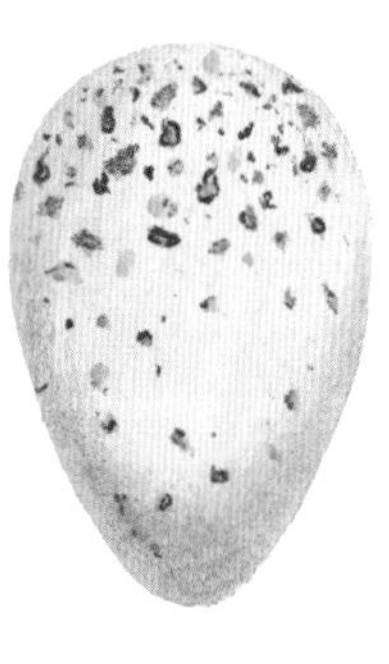

Nesting on the ground, as many birds do, means that eggs and chicks are very vulnerable to both mammalian ground predators and aerial avian ones. Such species often have rudimentary nests; in the Northern Lapwing's case it is just a scrape in the ground, and highly cryptic eggs to avoid detection and losses to predators. Northern Lapwings guard their eggs and chicks by 'mobbing' predators which approach their nesting areas. When a predator, such as a Carrion Crow, comes close to the group of nesting Lapwing many of the adult birds will take to the air and attack it by repeatedly diving at it in order to drive it away. Similar behaviour is directed at ground predators such as stoats or foxes, which they dive-bomb threateningly.

The Lapwing is a wading bird found across a swathe of northern Asia and Europe, but has declined over much of its European range because of the impacts of intensification of agriculture. (Both adult birds and eggs have also been a favoured food for humans.) This means that where Lapwings nest in loose groups, the overall number of birds in the colonies has fallen.

A single pair of Lapwings can be very brave in attacking predators, but it is obvious to the observer that the more birds there are the more effective they are at deterring the hunters. Ground predators duck their heads when a plunging Lapwing flies close to them. If numerous birds are doing this repeatedly and from different angles, the predator becomes distracted and is more concerned with its own well-being than continuing to search for nests or chicks. Scientific studies confirm this, and show that Lapwings nesting in larger groups suffer lower losses to predators.

RED-BILLED QUELEA

Quelea quelea

ABOVE
Red-billed Quelea are weaver birds, related to the House Sparrow (*Passer domesticus*) but about half its size. A huge flock of Red-billed Quelea darkening the sky and swirling in tight formation is an impressive and awe-inspiring sight, although it becomes a frightening spectacle for farmers: large numbers can ravage crops.

The most numerous wild-living bird on Earth (there are many more chickens), with an estimated population of over 3 billion, the Red-billed Quelea travels the African grasslands, following the rains which allow the grasses to grow and seed, on which it feeds. Flocks of over 40 million have been recorded, and colonies of over a million nests are not uncommon.

Red-billed Quelea build their nests by weaving grasses together. The males begin the nest-building and create an open oval structure, but when a female chooses to pair with a male the two birds complete the nest together, making it fully enclosed, with a narrow entrance. The nests hang from branches out of reach of most predators and are crowded together on bushes and trees, which can look as if they have been decorated with baubles.

Females lay 2–4 eggs and the incubation period is short, as is the time the young spend in the nest. This is probably an adaptation to the limited nature of the seasonal food resource, and the intense competition for it because of the large colony size. But Red-billed Quelea are also adapted to multiple nesting attempts, often large distances apart. Birds nesting in Ethiopia one June were found nesting again 100 days later about 1,000 km (620 miles) away.

Red-billed Quelea will readily turn from wild grasses to planted crops such as sorghum, millet and rice. A million Red-billed Quelea can destroy the crops of a small village in a single day and they are sometimes called 'Locust Birds'. Attempts have been made to control and reduce their numbers, and colonies are sometimes dynamited by local farmers or sprayed from the air with an insecticide that is highly toxic to birds. Despite these attempts, the Red-billed Quelea is as abundant as ever, and may even be increasing in numbers.

ROOK

Corvus frugilegus

The Rook closely resembles its relative the Carrion Crow (*Corvus corone*) in shape and size, and in its black plumage. But they are not completely similar. Children were once taught that 'a flock of crows are rooks, and a single rook is a crow' because a major difference between the two species is their sociability. This is nowhere more obvious than in the fact that the Carrion Crow nests in pairs on separate territories, whereas the Rook nests in tree-top cities called rookeries. So characteristic a feature is this, that the term 'rookery' is often used for other colonial-nesting species such as penguins.

The rookeries are traditional sites, and individual nests survive from year to year but need renovation each spring. The birds build their nests with twigs almost always collected in the trees rather than from the ground, though many are stolen from neighbouring nests when the opportunity arises.

Rooks are found across northern Europe and much of Asia. They begin to occupy their rookeries in early spring, often before the trees are in leaf, so their presence is obvious to all and their gathering at the traditional sites is a sign of the coming of longer days. The eggs, usually 3–5, are laid in late February or early March.

The average rookery in the UK consists of around 30 nests, but the largest ever counted, in Scotland, held over two thousand nests. Quite why Rooks are so social is still unknown, but their traditional rookeries, and their apparent keenness to get on with nesting in the early days of spring, have endeared them to the rural communities with which they share the countryside.

ABOVE
Adult Rooks (below) have a bare patch at the base of their bills, which young Rooks (above) develop during their first winter.

OPPOSITE
The Rooks Have Come Back by Russian Alexei Kondratyevich Savrasov (1871) captures the Rooks' early nesting season and the hubbub at a colony.

BELOW
Rooks lay three to five eggs, which hatch after fifteen days.

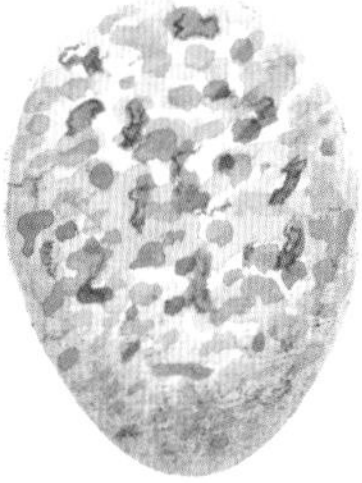

The Puffin

ATLANTIC PUFFIN

Fratercula arctica

ABOVE
Puffins nest below ground in burrows, often at the top of cliffs. Sometimes they take over old rabbit burrows, as shown in this illustration; the puffins appear to be driving out the original occupants.

OPPOSITE
The Atlantic Puffin is immediately recognizable by its upright stance, black-and-white plumage and characteristic bill. The brightly coloured bill is used in courtship displays, but it is also a tool for digging burrows and carrying fish back to the chicks, as well as a weapon for fighting other puffins.

The Atlantic Puffin is a member of the auk family of seabirds, which breed only in the northern hemisphere. But it is an unusual one in that it nests in burrows, which it digs in the soil of cliff tops overlooking the sea, rather than on the cliffs themselves. This limits the locations where the puffins can nest and also means that favoured sites often take the form of large colonies of thousands of pairs; although the Westmann Islands of Iceland harbour over 1 million pairs. Such nesting habits make the puffins' nests, eggs and chicks potentially vulnerable to mammalian predators such as rats or cats, and so the nesting colonies are found on some of the most remote, predator-free islands.

This fish-eating seabird is instantly recognizable because of its large, brightly coloured bill, though in fact the bill shrinks in size and loses much of its colour during the non-breeding season. The bill is used in courtship displays, for fighting other puffins, excavating the nest burrow and carrying fish back to the chicks. In the colonies adults also bond together by touching their bills, known as 'billing', in behaviours that cannot but remind the human observer of kissing.

Outside the breeding season, the Atlantic Puffin is found far out at sea over much of the North Atlantic Ocean. It feeds on a variety of fish including herring and sandeels, catching them by diving from the surface and then using its wings, almost as in flight, to propel itself underwater in pursuit of its prey. When taking a catch back to the colony to feed chicks, several fish may be lined up in the bill for the return journey.

The same nest burrow is used from year to year, and sometimes the burrows of rabbits are taken over. In spring, the returning birds (it is thought that the members of a pair reform their bond by meeting at the burrow each spring, rather than staying together through the winter months) refurbish their nest-site. A single white egg is laid in the nest chamber, and the parents share the

incubation duties. After around six weeks' incubation the chick hatches and the parents start bringing beak-loads of fish to feed it.

Even though puffins choose nesting locations that are free of mammalian predators, their young are still vulnerable to avian ones, with the main threats being large gulls and skuas. So the puffin chicks stay hidden in their burrows until they fledge, at around six weeks. They then launch themselves off the cliff tops at night and use their wings in their first, unsteady flight to settle on the sea beneath the colony and paddle away. By daybreak they are over a mile away.

While provisioning their chicks, adults feed at a distance of up to 100 km (60 miles) from the colony. When returning with their beaks full of fish the adults must also run the gauntlet of large gulls and skuas, which try to steal their catch by chasing them and attacking them in the air. Often parents gather in small groups and arrive at the colony together to minimize the risk of being robbed by these avian pirates. Human predators also prized Atlantic Puffins, both as food and for their feathers. Birds are still caught and eaten, under licence, in Iceland and the Faroe Islands. Nets and nooses are the main techniques used to snare them as they arrive back at the colonies after spending time at sea.

Where colonies have been abandoned, model Atlantic Puffins have been deployed to try to attract birds to re-establish the old nesting grounds, coupled with the reintroduction of young Atlantic Puffins from other colonies. These attempts have sometimes worked successfully – so successfully in fact that live birds have even displayed and 'billed' with the wooden decoys.

ABOVE
Atlantic Puffins spend most of their lives at sea, coming to land for just a few months to lay their eggs and tend their young. A single egg is laid, and the parents share incubation and feeding duties. From laying the egg to the young leaving the nest will tie the parents to land for around 12 weeks.

OPPOSITE
Puffins pursue their prey by diving from the surface and propelling themselves underwater using their wings. They catch beakfuls of fish to take back to their young.

2/5

Litho. W. Greve, Berlin.

PUFFIN.

Fratacula artica *(Linn.)*

EMPEROR PENGUIN

Aptenodytes forsteri

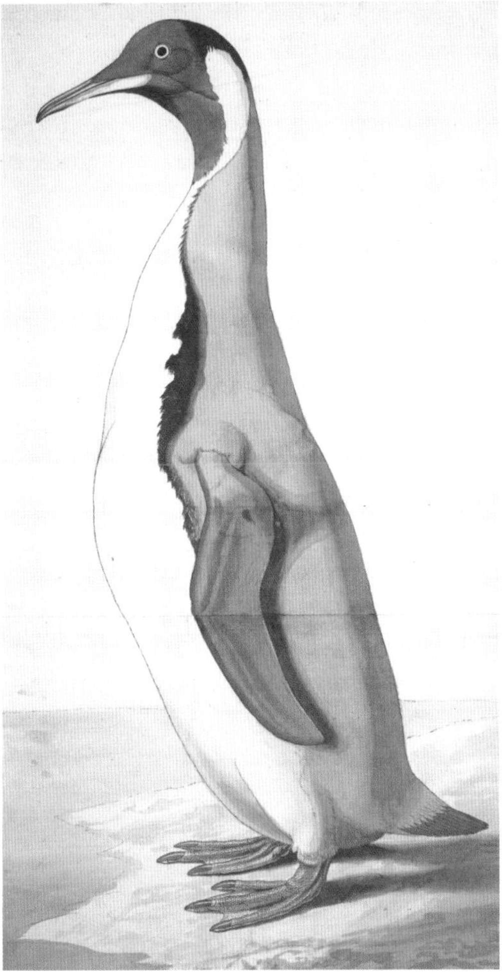

ABOVE
The Emperor Penguin is the tallest and heaviest of the world's penguins. Like all the others it is flightless, with short wings that it uses to great effect in swimming under the waves to catch its prey.

OPPOSITE
Emperor Penguins nest in huge colonies called rookeries. The colonies are established far from the sea, but as summer approaches, the ice melts and the young penguins have a short walk to the water.

In possibly one of the most inhospitable 'nesting' sites in the world, male Emperor Penguins gather together in large numbers on frozen ice, tens of kilometres from the open sea, to spend the Antarctic winter incubating their single eggs. They build no nest – there is no nesting material available on the icy waste of the Antarctic continent – but balance the eggs on top of their feet and in the warmth under their feathers in a brood pouch. Hundreds, or even thousands, of males will congregate at colonies and huddle together to try to conserve heat as incubation takes place in the coldest part of the Antarctic winter. It is thought that they take turns at being in the centre and at the edge of the flock, so that all get their share of sheltered and exposed positions.

Both male and female Emperor Penguins travel to the colony sites in the southern autumn (April), with the members of a pair searching each other out and re-establishing their pair bond. But after the females lay their eggs, and the males have begun incubating them on their feet, the females depart to replenish their body reserves for the time later in the season when both parents tend the young. For two months (June and July) the males endure the coldest temperatures on Earth, down to minus 40°C (-40°F) and winds of almost 144 kph (90 mph) – all without feeding and so relying on their fat stores. During incubation the eggs are kept warm, at temperatures of 38°C (100°F), while on the other side of the fold of fat, skin and feathers that comprise the brood pouch, the air temperature can be minus 38°C (-36°F), with an added high wind-chill factor too.

When the birds form their colonies or gatherings they are often as much as 160 km (100 miles) from the sea, but as the ice melts during the spring the sea approaches closer and closer so that the young have only a relatively short distance to travel to take their first swim in the Antarctic Ocean. Soon after the young hatch, the females reappear to feed them and the males are released from their duties for a period of around 3–4 weeks, when they replenish their food reserves. After that, the males come back to the colony and the two parents take it in turns, as one guards and tends their chick in the colony, while the other walks to the sea, feeds and returns to regurgitate food for the young one. While small, each chick is still protected in the brood pouch just as the egg was.

It seems an arduous process and one that tests both parents to the limit. If one parent does not return to the colony in time to take its turn in tending the chick, then the other parent there will desert the chick as its body reserves get dangerously low and leave for the ocean in order to feed. Deserted chicks soon perish of cold and starvation.

Three British explorers and scientists, Apsley Cherry-Garrard, 'Birdie' Bowers and Edward Wilson, on the ill-fated Scott *Terra Nova* expedition to the Antarctic, made what Cherry-Garrard described as *The Worst Journey in the World* (the title of his book about it) to secure three Emperor Penguin eggs for science. They set off on what would be a 35-day tramp through the total darkness of the Antarctic winter, braving gales, crevasses and the coldest temperatures on Earth to collect eggs that they hoped would illuminate the relationship between birds and reptiles. They suffered greatly and succeeded in their quest, but the specimens were of far less scientific interest than they had hoped.

At first it might appear odd that the parent penguins walk such great distances from the edge of the pack ice to breed and form their colonies, but it makes more sense for the adults to make this long journey once, rather than having to waddle away from the approaching sea edge as the ice gradually melts in spring during incubation or when the chicks have hatched. The chicks are vulnerable to just a few avian predators, mostly Antarctic Skuas, and the remoteness of the colonies means that the risk is low.

Once fully grown, the young Emperor Penguins will take to the sea and feed on fish, crustaceans and squid for at least three years before becoming sexually mature. They first breed at between three and eight years old, when they will make the long walk to the traditional colonies, far from the sea, to raise their own young, and the cycle repeats.

OPPOSITE
Incubation of the single egg is done by the males, who balance the egg on their feet and tuck it into a brood pouch as they sit out the Antarctic winter and the coldest temperatures on Earth. No other birds incubate their eggs in such hostile conditions.

OVERLEAF
Edward Wilson's drawing of an Emperor Penguin. Wilson and his companions walked for 35 days through freezing conditions to visit an Emperor Penguin colony, to collect eggs for science.

Wolf del et lith.

Printed by Hullmandel & Walton.

APTENODYTES FORSTERI. G.R.Gray.

E.A.W.

Emperor Penguin
farthest East. Jan. 31 '02

Useful to Us

Red Junglefowl • Turkey • Common Pheasant
Fishing Cormorants • Canary • Greater Honeyguide
Domestic Pigeon

ABOVE Turkey.

We may often be sentimental about birds, and appreciate their beautiful songs, mastery of the air and magnificent plumage, but we tend to forget this when it comes to eating them. Birds' eggs and meat are elements of the diets of just about all human cultures. As we go to the shops or markets to buy our food, some of our most popular options, often close together on the shelves, are derived from birds which originate in very different parts of the world.

The domestic hen or chicken is related to the Red Junglefowl of the forests of Asia. Thanks to our liking for its eggs and flesh, this is now by far the commonest bird in the world. Sometimes, for variety and especially for certain festivals and at particular times of year, the shopper may choose turkey. The Wild Turkey is a native American species of woodland and lives far from the Red Junglefowl. As consumers, we are likely to be familiar with both these birds, despite never having seen either in the wild or knowing exactly where they come from originally.

On the other hand, the Common Pheasant is a species that might be more familiar in the wild than in the shops. Although it is a bird of open forested country in Asia, it is now found on many other continents. It is mainly prized as a gamebird and has been introduced into several parts of the world so that it can be hunted and then eaten. Originally also domesticated for its eggs and flesh, the pigeon remains a companion of humankind for other characteristics. Its ability to travel long distances and a homing instinct to return to its coop have both proved useful and become the basis for a sport.

There are some birds that, rather than forming part of our diet, actually help us to find food. Cormorants of two species have been used in several parts of the world, but especially China and Japan, to do our hunting for us. Domesticated cormorants were employed to catch fish for fishermen, a practice that continues as a tourist attraction to this day. Another form of help in finding food for us is provided by the Greater Honeyguide, which leads our species to wild bees' nests in the savannah of Africa.

Our last example is not a bird that feeds us, but it has in the past saved many lives. The Canary, because of its greater susceptibility and sensitivity to gases noxious to us, was a valuable early warning device for miners working below ground.

RED JUNGLEFOWL

Gallus gallus

ABOVE
The familiar domesticated chicken, the world's most numerous bird, is derived from two forest-dwelling species of junglefowl.

OPPOSITE
The Red Junglefowl, here painted by an Indian artist, is the main ancestor of the domestic chicken and lives in forests of Southeast Asia. Domestication has brought chickens to every continent, either as a farm animal or as a source of meat.

It may be surprising to learn that a bird that lives wild in the forests of Asia from India to Indonesia, the Red Junglefowl, is now the commonest bird in the world by far, at least in its domesticated form of the chicken. There are something like 20 billion chickens – more than two for each of the world's human population – vastly outstripping the wild population of this or indeed any other species of bird. Only a small proportion of people will ever see a wild Red Junglefowl, but many will come across a domestic chicken either scratching around for food like its wild cousin, or more probably as a source of meat and eggs, since chickens now form such a ubiquitous part of the human diet.

The wild bird is highly sexually dimorphic – that is, the male and female are noticeably different in appearance. Male junglefowls are quite showy, having red fleshy wattles and a 'comb' on their heads, with a 'shawl' of orange and yellow feathers spreading over the chest, throat and nape. The smaller females, who do all the incubation of the eggs and care for the young chicks, are brown, with a less gaudy version of the male's yellow shawl.

Outside the breeding season, the birds live in small flocks. Males look out for predators and utter alarm calls when one is spotted; different calls are used for ground and aerial predators. In the breeding season, dominant males defend territories occupied by several females, while subordinate males form bachelor flocks. Males fight to establish a 'pecking order' of dominance – it was studies of chickens that led to that term – and they have spurs on their legs with which they fight when disputes get serious. This behaviour is the basis for cock-fighting, which is another use to which we have put the familiar version of this jungle bird. In fact it's possibly the reason why they were first domesticated.

OPPOSITE
The Grey Junglefowl (*Gallus sonneratii*) contributed some genes to the domestic chicken, although the Red and Grey Junglefowl rarely hybridize either in nature or in captivity.

BELOW
The domestic chicken occurs in a range of colours and different shapes and sizes, but is familiar to people all over the world.

Domestication occurred at least 7,000 years ago in both China and India, and since then the bird has spread with human travellers across the world. Despite our great reliance on them, however, we do not always treat them well – most chickens these days see little except the inside of a shed, packed together in very restricted cages. Before such factory farming, and still in many places, chickens are also reared on a small scale – the birds do well on scraps of waste from cooking and agriculture, thus performing a useful task of clearing up unwanted vegetable matter and turning it into edible meat or eggs. Chicken meat is low in cholesterol and fat compared with red mammalian meats; it is also incredibly versatile and has been adapted to many different cuisines.

Perhaps because of its long and close association with humans, the humble chicken has been the subject of intensive breeding programmes, whether selecting for producing more meat or eggs or purely for ornamentation, and there are now perhaps hundreds of different breeds, some of them very fancy. Chickens, like pigeons, were also used by Charles Darwin when developing his ideas about evolution by natural selection.

The cockerel or rooster is one of the twelve animals that are symbols of the Zodiac in the Chinese calendar, and people born under this sign are thought to be hardworking, honest and flamboyant. The chicken is also significant in the religions of Japan and India, but has penetrated religion and culture far beyond the lands of its origin, and has often been associated with divinatory and sacrificial rituals. In Judaism a cockerel is sacrificed on the eve of the Day of Atonement. In Christianity Pope Gregory I in the sixth century declared the cockerel as the symbol of Christianity, and it still sits atop many church towers as a weather vane. Christians are reminded of the time when St Peter denied knowing Christ three times before the cockerel crowed.

The crowing call of the male, used in its home to attract females in the thick undergrowth, now resounds round the world and is the first thing that many people hear on waking, as the cockerel heralds the dawn.

TURKEY

Meleagris gallopavo

The Wild Turkey is one of the largest – and in some ways most spectacular – birds of North America, and is found widely across the USA, northern Mexico and just into southern Canada. It is primarily a bird of broad-leaved forests, although it can be seen also in open ground. The males (typically 5–11 kg or 11–24 lb) are much larger than the females (typically 2.5–5.4 kg or 5.5–11.9 lb), but the record largest male ever hunted, in 2015 in Kentucky, USA, weighed in at over 17 kg (37 lb).

Turkeys were domesticated over 2,000 years ago in Mexico, and perhaps their earliest use in the American Southwest was ceremonial, with their bones turned into whistles and their feathers used to make textiles, blankets, masks and headdresses. Now, of course, turkey is one of the commonest forms of farmed poultry across the world, though it still lags a long way behind the domesticated chicken as a source of food. And unlike chicken, turkeys are mostly produced for their meat rather than their eggs.

The social system of the Wild Turkey is one of polygyny – males ('toms' or 'gobblers') will mate with several females ('hens') and play no part in the parental care of the young. Males 'strut' to display, spreading their tails and wings, hissing and gobbling, and dragging their wings on the ground. Some work as a team – two (sometimes more) males will stay close together, with one of the males doing most of the performance. The pairs of males have been shown to be brothers displaying together. Dominant males in the pairs are thought to mate with six times as many females as males who display alone, so it clearly pays to have a companion when displaying.

OPPOSITE
Turkey by the Mughal painter Ustad Mansur (*c.* 1612). This North American woodland species was taken to Asia and Europe from the sixteenth century onwards.

Benjamin Franklin, one of the Founding Fathers of the United States, was a great admirer of the Wild Turkey. He commented in a letter to his daughter that the Bald Eagle depicted on the American Seal looked more like a turkey than an eagle, disparaging the Bald Eagle as a bird and praising the Wild Turkey: 'For in Truth the Turky [*sic*] is in Comparison a much more

ABOVE
Would the Wild Turkey have been a better choice of national bird for the USA? That was the view of Benjamin Franklin, who preferred its character to that of the Bald Eagle.

respectable Bird, and withal a true original Native of America.... He is besides a Bird of Courage, and would not hesitate to attack a Grenadier of the British Guards, who should presume to invade his Farm Yard with a red Coat on.'

Notwithstanding Franklin's admiration for the Wild Turkey, Americans, since the mid-nineteenth century, have eaten its domesticated relative on Thanksgiving Day to commemorate the first feast between the Wampanoag Native Americans and European colonists at Plymouth Colony in 1621. At that feast a wide variety of food was eaten, and some believe that the birds on the menu would as likely have been Heath Hens, a now extinct woodland grouse. Not bothered by such doubts over the authenticity of seventeenth-century bird identification, Americans and Canadians sit down today to what is now a traditional meal of roast turkey on the fourth Thursday of November. Around a third of turkeys eaten in North America are consumed in the month that starts with Thanksgiving and ends with Christmas.

Roast Turkey is also a customary Christmas meal in many other parts of the world, having spread from Europe, where it is suggested that King Henry VIII of England may have had the first Christmas dinner of turkey. Around 5.5 million tonnes of turkey meat are produced worldwide annually, over half of it in the USA, which amounts to about 2,500 times the combined weight of all the free-living Wild Turkeys in the world today.

The wild bird was also hunted as a food source, and by the early twentieth century numbers had declined. Fortunately, breeding and management programmes have reversed this trend, so that the Wild Turkey is now once again strutting through the forests of North America as it has done for thousands of years.

COMMON PHEASANT

Phasianus colchicus

Another native of Asia that has been introduced widely around the world, the Common Pheasant's original homeland stretches from the eastern shores of the Black Sea to China and Thailand, where it lives in lightly forested land. The bird's Latin name, and the English word 'pheasant', come from the ancient town of Phasis, near the modern port of Poti in western Georgia.

Today, because of interbreeding of different varieties over the centuries, numerous colour varieties of the captive-bred gamebird can be found, but all have a long tail and most a green head with red eye-wattles. The body is a mixture of speckled gold and brown, and many have a white neck collar. Males fight among themselves in the spring mating season and defend harems of females. The females, since they generally do all the rearing of the chicks, are smaller and much less conspicuous – they are brown and about 25 per cent lighter in weight than the males, and have shorter tails.

ABOVE
The Common Pheasant is a gamebird which is reared and released into the wild for shooting. The male (below) is about a third heavier than the female (above) and has much brighter plumage.

The main reasons for the introduction of pheasants to many parts of the world were first as food, and, more recently, for sport. The Romans were responsible for the spread of the bird across their empire in Europe. The late fourth- or early fifth-century AD collection of recipes attributed to Apicius includes one for pheasant dumplings. In more recent centuries the birds were taken and released into the wild in North America, Australia and New Zealand.

In the USA, particularly the Midwest, Pheasant shooting is a favoured form of hunting. A party of shooters walk through suitable habitat (woodland edges, brush, farmland), usually with dogs, and shoot at any Pheasants that are flushed out. Some breeds of dogs used, such as retrievers, flush the birds, while others, such as pointers, just locate and mark the birds, which are

flushed by the shooters themselves when they are ready to shoot. When flushed from cover, Pheasants have a rapid and towering flight, often accompanied by a harsh call of alarm and loud whirring of wings. In South Dakota, where the non-native Common Pheasant is the official state bird, well over a million Common Pheasant are shot each year. A high proportion of these are wild-bred feral birds rather than being managed and reared for the purpose.

This pattern of hunting was the norm in the UK until the mid-eighteenth century. With the invention of the breach-loading shotgun it then became possible to reload much more quickly, so that more birds could be shot. This in turn allowed the evolution of driven shooting, with a line of 'beaters' moving through suitable habitat flushing the pheasants forward towards a stationary line of 'guns'. The pheasants break cover and fly over the line of guns. A high steepling pheasant takes a skilled shot to bring it down, but even so vast numbers are killed. There are many astonishing records: for instance, on 18 December 1913, in Buckinghamshire, 3,937 Pheasants were shot by a party of seven guns including King George V, a record 'bag' of Pheasants for one day. Pheasants are normally hung for days before plucking and cooking, as with most other game meat, and are usually eaten roasted.

About 45 million Common Pheasants are reared in captivity and released into the UK's woods ahead of the shooting season established by the Game Act of 1831, of which about 25 million are shot each year. The Common Pheasant, though not native to the UK, is thus, for most of the year, by far the commonest bird in the British countryside.

ABOVE
The Common Pheasant is an attractive and familiar sight in the countryside in many parts of the world. Though it originated in Asia, it now features far from its homeland in sport, art and cuisine.

OPPOSITE
Common Pheasants have been selectively bred for different colour forms. Not all males have the white neck colour, and some have generally green plumage, although the standard form is brown and gold.

FISHING CORMORANTS

Phalacrocorax carbo &
Phalacrocorax capillatus

ABOVE
There are around 40 species of cormorant and most are marine, feeding on fish and nesting on cliffs or islands to avoid mammalian predators. Almost all have an entirely black plumage, which has led to an association with evil.

OPPOSITE
Swallowing a large fish whole can seem a difficult task for a cormorant and may take several attempts, but after some juggling the bird succeeds. This behaviour has given cormorants a reputation for gluttony and has sometimes made them unpopular with fishermen.

There are around 40 species of cormorant across the world, occupying every continent. All are fish-eaters, and this has sometimes led to conflicts with humankind in places where the birds visit commercial fish stocks. The cormorant, most particularly the Great Cormorant (*Phalacrocorax carbo*), is sometimes represented as an emblem of evil and gluttony, and it is easy to see how this unflattering depiction developed. Cormorants catch fish by diving underwater and then bring them to the surface to swallow them whole. But the bird must first position the fish correctly, as many have spiny fins that would make swallowing them a painful or difficult process if downed the wrong way. The swallowing of a large fish can take several attempts, with the bird seeming to juggle its catch in its bill. Suddenly the fish disappears – except sometimes it only gradually disappears. A large bulge is visible in the bird's long neck as the fish is slowly swallowed and consumed. A person behaving similarly at the dinner table would certainly be thought of as gluttonous.

The entirely black plumage of almost all cormorants also hasn't helped their reputation and the bird is associated, in European cultures at least, with evil. Satan, in Milton's *Paradise Lost*, took the form of a cormorant to spy on Adam and Eve: 'Thence up he flew, and on the Tree of Life, the middle Tree and highest there that grew, Sat like a Cormorant.' And so cormorants do not have a uniformly favourable character, though there are a few places – Greece, China and Japan in particular – where the cormorant's skills as a catcher of fish have led to a long partnership with coastal communities.

The Great Cormorant is found widely across Europe, Asia, Australasia and Africa, and has a toehold in North America. It is found in coastal waters,

ABOVE
Cormorants' feathers lack water resistance and it is not uncommon to see the birds out of water drying their plumage, often with their wings extended in a cruciform pose.

OPPOSITE
In Japan and China cormorants have been used by fishermen to catch fish for centuries. The birds are rewarded for their work with a share of the catch.

but also far inland, feeding on lakes and large rivers. In China and Greece, Great Cormorants have been tamed and kept in captivity by man for well over a thousand years. In Japan it is the slightly smaller Japanese Cormorant (*Phalacrocorax capillatus*) that is used in this way. These trained cormorants are taken out by fishermen and allowed to catch fish. A line tied around the cormorant's neck, and leading back to the boat, prevents it from either escaping or swallowing large fish (although small fish, of little interest to the fishermen, can be swallowed). The cormorants learn to bring captured fish back to their owners and are rewarded at the end of the fishing trip with a portion of their catch.

Fishing with the Japanese Cormorant (*umiu*, Sea Cormorant, in Japanese) takes place in thirteen cities in Japan; in Gifu on the Nagara River the practice has occurred, unbroken, for at least 1,300 years. Here, and in nearby Seki, on the same river, the cormorant fishing masters are employed by the emperor and known as Imperial Fishermen of the Royal Household Agency. What began as a means of catching fish for local consumption was transformed first into an industry as catches increased and now is a major tourist attraction.

Surprisingly, the feathers of the cormorant lack water resistance, despite its aquatic way of life. This is why cormorants are often seen perched on some convenient structure, natural or manmade, with their wings extended to dry their feathers. This stance has been incorporated into heraldry and in the Christian world is associated with the cross of Christ. And whereas in the minds of westerners the cormorant's ability to swallow fish whole evoked gluttony, in Japan it gave rise to a rather different interpretation. The phrase 'to swallow like a cormorant' was used to suggest that someone is gullible and has accepted a suggestion without thought or question.

Thus we seem to have mixed feelings about the cormorant. When they are catching food for themselves we are not always their greatest admirers, but when they catch fish for us we regard them as very special and useful birds.

CANARY

Serinus canaria

In the wild, the Canary is a small greenish-yellow, seed-eating finch which inhabits the Atlantic islands of the Canaries, the Azores and Madeira. There it is common in all habitats, and from sea level to mountaintops. But this species has also long been domesticated as a cage bird, because it is easy to keep, feed and breed, and the male has an attractive twittering song that we enjoy.

Through selective breeding, various colour forms of Canary have been developed (including white, black, brown and even red) and a wide range of conformations and songs now exist. Canary breeders have shows, involving competitions in which domestic canaries are compared and assessed, with prizes awarded to the owners and breeders of the best birds. Canary songs are judged by their variety, frequency, tone, carrying capacity and a whole range of other attributes in a 20-minute judging period.

Canaries have also been put to more practical use. They were kept in coal-mines to alert miners to the presence of poisonous gases such as methane and carbon monoxide, to which Canaries are more sensitive than people. The first signs that the birds were suffering would be a warning signal to the miners. This practice continued into the 20th century, and the phrase 'Canary in the coal-mine' is still used as a metaphor for any change in wild species' status that might indicate changes in ecological conditions that could affect human well-being too.

Despite a wild population restricted to small Atlantic islands and numbering only in the tens of thousands, this diminutive bird now has a worldwide presence thanks to our love of its song and for its ability to keep us safe.

ABOVE
Canaries have been prized as cage birds for centuries. Their twittering song, bright plumage and chirpy behaviour make them attractive household companions for us.

OPPOSITE
The Wild Canary has a yellow breast and head and brown streaked back and wings, but captive birds have been selectively bred into many colour forms.

GREATER HONEYGUIDE

Indicator indicator

ABOVE
The Greater Honeyguide uses the human love of honey to help obtain food for itself, by leading local people to a bees' nest and feeding on the remains after the honey has been extracted. Despite their given name, none of the other related honeyguide species lead people to bees' nests, although they do feed on them.

A dark grey-brown bird, found mainly in open woodland in sub-Saharan Africa, the Greater Honeyguide feeds primarily on the beeswax, larvae, eggs and pupae of bees. While it might sound a rather risky preference, the Honeyguide is able to do this by entering bee colonies early in the morning when the bees are still too cold to be active, or through feeding on abandoned bee colonies or on ones that other predators, such as Honey Badgers, have already ransacked. But remarkably, this bird also deliberately seeks human assistance to obtain its food. The Greater Honeyguide's scientific name reinforces the message of its English version – it indicates the locations of bees' nests to honey-hunters.

In this unusual partnership, the Greater Honeyguide will fly up to local people and attract their attention by calling and then fly in the direction of a bees' nest; it will call and lead again in stages. The human honey-hunter breaks open the nest to collect the honey, and the bird gains from being able to scavenge on the wax and other remains of the nest. Ironically, given its name, it largely ignores the honey.

It is not now generally thought (though it is disputed) that Greater Honeyguides lead non-human predators, particularly the Honey Badger, to bees' nests. If this is the case, it would mean the human–Honeyguide symbiotic relationship has evolved over many thousands of years, since the time of early humans. Today the behaviour may be dying out among Greater Honeyguides that live near urban areas, where the human population is high but few people are interested in nests of wild bees since other sources of sweeteners are more easily available.

DOMESTIC PIGEON

Columba livia

ABOVE
The Rock Dove is the wild species from which domestic and feral town pigeons are derived. It has a dark wing bar and a white rump and usually nests on cliffs.

A bird with a long association with humans, and nowadays loved and reviled in equal measure, the domesticated pigeon derives from the wild species of Rock Dove, found in North Africa, the Middle East and parts of coastal Europe. It is thought that pigeons were first domesticated as settled agriculture developed in Neolithic times in the Fertile Crescent of the Middle East.

Initially of course we would have been interested in the pigeon for its meat and its eggs. Wide-ranging in their tastes for food, pigeons are relatively easy to feed and breed in captivity. Adult pigeons regurgitate food in the form of a digested paste to their very young chicks (squabs), which means that if the adults are well fed at this time they are likely to breed well and produce more chicks. The young of other species often have more specific dietary needs and are thus more difficult to rear.

Once pigeons were domesticated for food, we then discovered many other beneficial things about them. Their ability to find their way back to their sites of captivity, and their willingness to do so, was soon exploited for carrying messages. Homing pigeons could be taken on a journey and later released with messages tied to their legs, in the knowledge that the bird would reliably fly to its 'home' where the message could be retrieved and read. Billions of migratory birds have the ability to navigate long distances back to their nesting and feeding sites, but the fact that a domesticated, captive bird will do the same both endeared the homing pigeon to us and made it useful. The ancient Greeks and Romans, as well as many other cultures through history and in different regions, have recognized and exploited the pigeon's capacity to do this.

Warfare is one situation in which pigeons have played a particularly significant role. The message announcing the outcome of the Battle of Waterloo was

BELOW
Due to their speed and excellent homing ability, pigeons have been used to carry messages 'back home' from travellers, boats at sea, besieged cities and armies in foreign parts. 'Pigeon post' has usually involved a folded message tied to the bird's leg or secreted in a small sack on the bird's back. This use goes back at least 2,000 years.

OPPOSITE
Originally pigeons would have been domesticated for their meat and eggs, but over the years they have also been kept for their beauty: their soft cooing helped to create a pleasant, soothing ambience, while selective breeding produced eccentric and extreme plumages and behaviours for sheer delight.

OVERLEAF
The buildings of our cities bear some resemblance to the cliffs on which wild Rock Doves nest. The pigeons find some 'natural' food in our cities (such as seeds), but their success is built on their ability to scavenge our leftovers.

said to have been delivered to England by a pigeon. As well as delivering reports of the battles, they were carried on British aircraft during the Second World War and were used by the American army, including in Guadalcanal. They have formed essential lines of communication when other means were not possible: in the Franco-Prussian war (1870–71) a 'pigeon post' was established to get messages into and out of the besieged city of Paris. Pigeons have been recipients of the Dickin Medal in the UK (often known as the 'animals' Victoria Cross') and the French Croix de Guerre in recognition of their gallantry. And more generally before the advent of electronic communication, messenger pigeons were of great importance. Reuters news agency transmitted stock market information by messenger pigeon before cable links were established.

The homing ability of pigeons has also led to the sport of pigeon racing, in which numerous pigeons are released by their owners at a set location far from their homes; their arrival back at their loft is recorded and prizes awarded to the fastest. This sport arose in Belgium in the mid-nineteenth century and birds have been selectively bred for their speed and homing ability. Average speeds of over 145 kph (90 mph) have been measured for flights of over 640 km (400 miles), and they can cover huge distances – one pigeon called Per Ardua flew over 1,610 km (1,000 miles) from Gibraltar to Kent in southern England.

Pigeons have also been kept for their beauty. The cooing sound of pigeons is soothing and over many years selective breeding has produced ornamental pigeons with striking, even rather odd-looking, plumages and behaviours, simply for our delight. The astonishing diversity of appearances and details of pigeons, all descended from the same original species, fascinated Charles Darwin, and this was another bird that was of great importance in his work on natural selection.

The pigeon was once a very common adjunct to our domestic lives, but as technology and farming have progressed, its usefulness, except in aesthetic terms, has been greatly reduced. But pigeons still live in close association with us in the form of feral pigeons, which are from the same genetic stock as the Rock Dove and domesticated pigeons. Wild Rock Doves nest on cliffs, and modern buildings resemble those habitats; in addition, the pigeon's adaptability has allowed it to survive well on the scraps of food that it can find in modern cities. Feral pigeons, often in large numbers, are today regarded as a problem by many urban dwellers and city administrations.

Stab cromolit O. Dressler Milano

omba livia. Br.

Tav. 80.

Threatened & Extinct

Giant Moas • Dodo • Passenger Pigeon • Chatham Island Black Robin Huia • Ivory-billed Woodpecker • Lear's Macaw • Spoon-billed Sandpiper • California Condor • Great Auk

ABOVE Dodo.

About one in ten of the world's 10,000 bird species are currently regarded as under threat of extinction; unfortunately many have already succumbed. The giant moas of New Zealand were large – very large – flightless birds that were hunted to extinction by the human colonists of New Zealand centuries ago. We know little about them and can only try to imagine what it would be like to live in a world in which 3.5-m (11½-ft) high birds browsed the trees.

Being unable to fly may have been a disadvantage. The Dodo of Mauritius was another flightless bird that suffered extinction when its island was colonized by us, and the rats and cats we brought with us. A few accounts of the bird have come down to us, plus some paintings and badly stuffed specimens, and so, again, we can only try to imagine a world where this bird was not 'as dead as a Dodo'. Also flightless, the Great Auk, too, was hunted to extinction. This penguin-like seabird lived in the North Atlantic and was last seen off the Grand Banks of Newfoundland, Canada, in 1852.

One of the most amazing and famous avian extinctions was that of the Passenger Pigeon, a bird of the deciduous forests of North America. This species could darken the skies in flocks of billions, but the last wild bird was seen in 1900 and the final individual died a lonely death in Cincinnati Zoo in 1914. The speed of extinction of what was most probably the most numerous bird the world has ever seen is a cautionary story of how even the commonest species may be vulnerable.

Surprisingly, it is not always easy to know when, or even whether, a species is extinct. Perhaps individuals will be discovered in the future – how long should we wait before assuming none are still alive? Two such species are the Huia of New Zealand and the Ivory-billed Woodpecker of the USA – both have been assumed to be extinct before and have then reappeared; but hope for another confirmed sighting of either is slipping away.

These examples have helped to galvanize conservation action to save other birds. There are many uplifting stories of conservationists rescuing species from the very brink, and others where action is under way, though for some the ultimate outcome is uncertain. The California Condor, Chatham Island Black Robin, Lear's Macaw and the Spoon-billed Sandpiper are all the subject of attempts to halt decline and boost numbers.

GIANT MOAS

Dinornis novaezealandiae
& Dinornis robustus

THE MOA OF NEW ZEALAND.
(DINORNIS GIGANTEUS.)
From a Specimen in the Canterbury Museum, N.Z.

ABOVE
Giant Moas were flightless birds, roughly twice the height of the average human and, at 250 kg (551 lb), more than twice as heavy as the present-day Ostrich (*Struthio camelus*).

OPPOSITE
These extraordinary birds lived in New Zealand and were driven to extinction many centuries ago by humans hunting them for food.

Several species of moa, all flightless, and all long extinct, once lived in New Zealand. Two of these remarkable birds, one on North Island, the other on South Island, were much larger than the others and were known as Giant Moas. The South Island Giant Moa was the largest bird ever to have lived on Earth, being over 3.5 m (11½ ft) in height and weighing around 250 kg (551 lb). As the biggest creature in New Zealand, it had experienced virtually no predators until the arrival of Polynesian settlers after AD 1250.

Known only from oral records of the Māori, skeletons and scant other remains, it was until recently thought that there were three species of Giant Moa, but DNA analysis of bones has now shown that what was believed to be the smallest species was in fact simply the males of the presumed largest (South Island) species. Females were 50 per cent taller than the males and weighed almost three times as much.

Moas were vegetarian and their long necks allowed them, giraffe-like, to browse the leaves from tall trees. Some were forest-dwellers, but they occupied all of New Zealand's habitats and were an abundant ecological presence. They nested in rocky outcrops and built nests of twigs. The males probably incubated the eggs, as this is the habit of their closest living relatives, the tinamous and many others of a group of flightless birds such as emus and cassowaries.

Such a big, flightless bird was a very large and appealing packet of protein for humans, so it is perhaps not surprising that the indigenous Māori eventually drove these species to extinction around three centuries before the arrival of Europeans. They hunted the birds by driving them into pits and also robbed their nests. At one site where the birds were processed, archaeologists found the bones of as many as 90,000 moas – evidence of the scale of killing.

MEGALAPTERYX HUTTONI

(One-Quarter Natural Size—*restored drawing from feathers and mummified remains*)

DURABLE PAPER COLOR

DODO

Raphus cucullatus

For a bird that was last seen alive in the late seventeenth century – 1662 is often taken as the year of extinction as this was the last accepted sighting – and for which no complete specimens survive, the Dodo perhaps seems remarkably familiar. But in fact very little is known for certain about this flightless relative of the pigeon that inhabited the island of Mauritius, in the Indian Ocean.

The Dodo was first encountered by Europeans in the early sixteenth century, when sailors began to visit the island, and some live birds were brought back to Europe. We also have incomplete skeletons, some written descriptions and paintings (many inaccurate). From these we know that Dodos were large birds of around 1 m (over 3 ft) in height, possibly weighing over 20 kg (44 lb), though this is much debated. Such a big bird was a tempting target for any sailor making landfall on Mauritius and wanting fresh meat as a change of diet.

The flightless nature of the Dodo, together with the fact that as the island was uninhabited it had not evolved a fear of man, meant that it was relatively easy to catch. But, as with other extinctions, there were probably many contributory factors to its loss, and direct over-hunting by humans may not have been the most important. Habitat destruction and the introduction of pigs, cats and rats, which plundered Dodo nests, also played a significant role in the species' extinction.

Dodos kept in captivity may not have been fed on the correct diet, it seems, and became overweight, and any reconstructions today have to be made from composite elements from different skeletons, so we're not even sure precisely what the Dodo looked like or what their natural behaviour was. Perhaps the most accurate depictions from live birds are a sketch by

ABOVE
Dating from the mid-seventeenth century, this is possibly one of the last illustrations made of a live Dodo. Painted by the Dutch artist Cornelis Saftleven, it is a charming connection with a bird that has been extinct for three and a half centuries.

OPPOSITE
The Dodo must have been an odd-looking bird and this engaging caricature certainly captures its comical features. However, our understanding of the bird's ecology and behaviour is almost completely lacking.

123

S

London, Published Dec.r 1st 1792 by F. P. Nodder & Co. No. 15 Brewer Street.

RIGHT
Standing at half the height of a grown man, and with a massive beak, the Dodo must have been an impressive bird when it roamed the island of Mauritius, in the Indian Ocean.

the Dutch artist Cornelis Saftleven (1638), possibly one of the last illustrations made of a live bird, and one attributed to the painter Ustad Mansur (*c.* 1625), of a Dodo in the collection of the Mughal emperor Jahangir. It seems the Dodo was, to some extent, indeed rather a comical-looking bird, with a distinctively shaped beak and a round body (though perhaps not as exaggerated as in some portraits).

Alice famously encountered the Dodo in Lewis Carroll's *Alice's Adventures in Wonderland*, through which it became better known thanks to the illustrations of John Tenniel. This is perhaps the image of the bird that is lodged most firmly in many people's minds.

Modern research, including DNA study of remains in Oxford University, is throwing new light on this intriguing and much-loved bird, which has the perhaps unfortunate distinction of now being a symbol of extinction.

PASSENGER PIGEON

Ectopistes migratorius

ABOVE
The Passenger Pigeon was once the world's commonest bird, but not a single one survives today. Overhunting by humans in the nineteenth century and loss of native habitat led to its extinction; the last one died in captivity in 1914.

The last Passenger Pigeon on Earth died between midday and 1.00 p.m. in a cage in Cincinnati Zoo on 1 September 1914. The bird was a female called Martha (after Martha Washington, the wife of the USA's first President). For around fourteen years before this, the only Passenger Pigeons in existence were captive ones in collections in North America. But just a few decades earlier, this had been the commonest bird the world had ever known, forming flocks more than a kilometre wide that took hours or days to pass by. The name Passenger Pigeon comes from the French *passager*, as it was recognized as a bird of passage.

The Passenger Pigeon was a forest bird, feeding on acorns, beechmast and chestnuts in the deciduous forests of eastern North America. The trees produced their seeds in unpredictable quantities from year to year and place to place, so Passenger Pigeons were nomadic, travelling in large flocks and nesting in huge colonies of millions, often tens of millions, of birds where food was abundant. Some flocks, seen by famous and respected ornithologists such as John James Audubon, Alexander Wilson and John Muir, were estimated to number billions of birds, and as they passed across the sun, the earth darkened as if during an eclipse. The sheer weight of roosting birds could break the thick limbs off mature trees, and frequently used roosting sites were covered in droppings to the depth of several feet. Perhaps two in five of all North American birds were Passenger Pigeons – but now every single one is gone.

Native Americans had harvested the colonies for centuries, but had rules about not disturbing them too much and only taking the well-grown young to eat (they also tasted best according to many reports). With the coming of the railroad and telegraph, a growing population of European Americans pillaged

RIGHT
A pair of Passenger Pigeons, with the more colourful male above and the female below. The last two Passenger Pigeons to survive were named George (died 1910) and Martha (died 1914) after the Washingtons, the first President and First Lady of the USA.

OPPOSITE
John James Audubon painted all known American birds in his book *Birds of America* (1827–38), including this plate of the Passenger Pigeon. Audubon saw a flock of Passenger Pigeons in Kentucky, which he estimated to number a billion birds.

nesting colonies by shooting, trapping, cutting down the trees to get at the nests and even setting trees alight to make the young birds jump to the ground. Millions of birds were sent back iced or salted to the cities of eastern America to be eaten in restaurants, often as pigeon pies.

Such plundering in the late nineteenth century, added to the loss of most of the native forest of the eastern USA by 1872, reduced the Passenger Pigeon population to levels where natural predation, with which the bird could cope when abundant, perhaps finished it off. Geneticists hope that by using DNA from museum specimens they might be able to recreate the Passenger Pigeon, but it seems unlikely that we will ever again witness the skies darken as they are filled with millions of Passenger Pigeons searching for beechmast.

Passenger Pigeon
Columba Migratoria

CHATHAM ISLAND BLACK ROBIN

Petroica traversi

All species of bird are remarkable in some way, although some may justifiably be regarded as more remarkable than others, but few individual birds could claim to be remarkable. However, one female Chatham Island Black Robin, named 'Old Blue', might merit that label, since the world's entire population of 250 Chatham Island Black Robins today is derived from that single female.

This species of sparrow-sized bird has always been rare, restricted to the remote Chatham Islands, over 640 km (400 miles) southeast of New Zealand, but in 1980 the population was reduced to only five individuals: four males and Old Blue. Over a century before, the bird had been wiped out on the main island in the group by introduced predators such as cats and rats, but clung on in small numbers on Little Mangere Island.

In 1976, when the population was already as low as seven individuals, all the birds were caught and transported from Little Mangere to Mangere, where 120,000 trees had been planted in order to provide better habitat for them. But the future of the species hinged on the productivity of that last surviving female, Old Blue, who was already eight years old in 1980 – an advanced age for this species. She was named from the colour of a leg ring used to identify her.

In order to increase the annual output of chicks from Old Blue, and subsequent breeding females, New Zealand conservationists removed their first clutch of eggs (usually two, sometimes one or three) and placed them in the nests of another species, the Tomtit, where they were incubated, hatched and raised. Female Chatham Island Robins whose eggs were removed in this way usually laid another clutch, which they incubated and reared themselves.

This intervention, and the egg-laying ability and longevity of the remarkable Old Blue, enabled the population to grow steadily over time. Some credit

ABOVE
All surviving Chatham Island Black Robins are the descendants of a single female named Old Blue by New Zealand conservationists. Never before has a species come so close to extinction and been brought back from the very brink.

ABOVE
A male Tomtit (*Petroica macrocephala*) feeds two Chatham Island Black Robin chicks. Cross-fostering was a risky but successful ploy in increasing the population of the Black Robin.

should also go to Old Yellow – Old Blue's mate. Old Blue lived to an age of 14 years, much longer than the average for Chatham Island Black Robins.

But the increase in numbers did not proceed completely smoothly: a proportion of the growing female population began to lay their eggs on the rim of their nests where they could not be incubated. This seemingly odd behaviour put the conservationists in a quandary: should they intervene and push these eggs into the bowl of the nest or let nature take its course? Because the species was at such a low ebb, they chose to intervene. But because this behaviour was genetically programmed, it spread through the population. Eventually, as the population grew and seemed to be at a safer level, egg-pushing interventions ceased and the behaviour decreased in incidence.

In the end, after many years, the Chatham Island Black Robin, although still only numbering around 250 individuals, and still vulnerable, is in a much more secure position than it was 35 years ago. Many thought that it couldn't be done, that the population would be far too inbred to survive. Perhaps in the long term this will prove to be the case, but the success so far shows that some species, perhaps particularly those with small world populations because they have always been restricted in range, can go through very narrow genetic bottlenecks and come out the other side with a good chance of survival.

HUIA

Heteralocha acutirostris

ABOVE
John Gould's 1830s depiction of the sexual dimorphism in bill size and shape in the Huia (female above, male below).

OPPOSITE
John Keulemans's late nineteenth-century illustration of a pair of Huia. The last confirmed sighting of a Huia was less than 20 years after this painting was published, although hopes remain that the species might be rediscovered in a remote New Zealand forest.

Once found only in the forests of New Zealand, the Huia was a songbird of great cultural significance to the indigenous Māori. Despite, or perhaps because of this, it is now extinct. The last reported sighting was in the early 1960s, although the last confirmed record was much earlier, in 1907.

The cause of its extinction was twofold – habitat loss of the forests on which it depended and the killing of the bird for its feathers and bill; or, at least, for the female birds' bills. For in this species, bills of males and females were of very different shapes – far more so than in any other bird. Males had a fairly ordinary-looking straight, sturdy bill, but the females had extremely long, down-curved bills. We assume the females used their bills for probing in rotten wood for insect larvae, but little is known of this bird's ecology for certain.

Did males and females forage together or did the extreme differences in bill shape and size mean they exploited the habitat in different ways? Did females forage in all-female groups and males in all-male groups? Unless there are a few Huia remaining hidden in the forests of New Zealand's North Island, and we could find and study them, we will never know the answer.

Huias had predominantly black plumage, with distinctive white tips to the tail feathers. They also had bright orange wattles at the base of the bill. The females' bills were used in Māori jewelry, and the feathers were highly prized and were worn as adornments and symbols of status – the Māori even created finely carved treasure boxes to keep them in. The Huia was already probably under pressure from hunting by the Māori, but when a Māori guide took a Huia feather from her hair and placed it in the hat band of the visiting Duke of York (later King George V) in 1901, the value and popularity of such feathers soared and hastened the bird's decline.

IVORY-BILLED WOODPECKER

Campephilus principalis

It is in fact difficult to prove that something does not exist, and there are many examples of species once thought to be extinct later being rediscovered. Such creatures are sometimes called 'Lazarus species', as they seem to have risen from the dead.

Generally now regarded as extinct, the Ivory-billed Woodpecker is, or was once, widespread in the forests of eastern North America and was, or is, one of the largest woodpeckers in the world (with a body length of 50 cm or 20 in. and a wingspan of 75 cm or 30 in.), the only larger one being the closely related Imperial Woodpecker (*Campephilus imperialis*) of western Mexico, which is also thought to be extinct. A very impressive bird, albeit rather similar in appearance to the common and widespread Pileated Woodpecker (*Dryocopus pileatus*), the Ivory-billed Woodpecker has the nickname of the Lord God Bird or Good God Bird, based on the exclamations by people when they first see this striking and large bird. It is also known as the Grail Bird, for it is avidly sought for, but on the whole now thought to be just a legend.

It may seem hard to believe that a very large woodpecker could exist in North America without people catching sight of it, but the preferred habitat of this species consisted of extensive tracts of woodland, sometimes flooded, and in areas of relatively low human population density. So if this magnificent bird does still survive, it is probably in small numbers in remote large forests. It would have been relatively widespread in southern states of the USA before tree-felling reduced the area of forest, and fragmented what remained. Such deforestation would also have led to the loss of dead trees and rotting timber, which were probably important for this bird's existence.

The Ivory-billed Woodpecker has been thought to be extinct more than once. In the 1920s ornithologists were surprised when a pair was seen, and promptly shot, in Florida. Then in the 1930s a bird shot in northern Louisiana dispelled the idea that the species was by then certainly extinct. In 1944 the last

OPPOSITE
John James Audubon was familiar with the Ivory-billed Woodpecker from his home in Kentucky, not far from where the Ohio River meets the Mississippi. He likened the bird's plumage to the 'boldest and noblest productions' of Van Dyck's pencil.

OPPOSITE
Does the double 'Tap! Tap!' of the bill of an Ivory-billed Woodpecker against an ancient tree still ring out in some river valley or swamp in the southern USA, or is this a sound that ceased decades ago? Opinions differ, and time will tell.

known female was seen, which led to the listing of the species as extinct five years later by conservation authorities.

Since then, reports of Ivory-bills in its former range have continued to emerge, but most of these are now discounted by the ornithological world. Anyone claiming to see an Ivory-billed Woodpecker today is likely to come under much questioning and face great scepticism, and for that reason some 'observers' have chosen to remain anonymous rather than experience such intense interrogation.

Hopes were raised in 2005 when a group of ornithologists from Cornell University published a paper in the journal *Science* entitled 'Ivory-billed Woodpecker ... persists in Continental North America'. The paper, with its optimistic title, suggested that a male Ivory-billed Woodpecker had been seen in the Big Woods of Arkansas. However, the evidence provided was regarded by many as falling far short of proof. Some commentators felt strongly that the reported sighting was simply of the common Pileated Woodpecker. Large amounts of time, money and effort have since been spent on searching for the Ivory-billed Woodpecker, but without producing any definite sightings with photographic evidence.

And so we remain unsure as to whether this intriguing bird is still sharing the Earth with us, or whether the last one perished in the 1940s, 1950s or perhaps much more recently. If it remains, somewhere in the remote forests of the southern USA, and the characteristic 'Tap! Tap!' rapping of its bill on tree trunks still echoes through some distant wood, then it must be in extremely low numbers. Will it rise from the dead once again, or is it actually long gone? What is certain is that people continue to be fascinated by the Ivory-billed Woodpecker and seem driven to search for it in the hope that it still survives, somewhere.

LEAR'S MACAW

Anodorhynchus leari

It was only in 1978 when two very small isolated wild populations were found in northeast Brazil that this large blue parrot was finally determined to be a distinct species in its own right, rather than a race of Hyacinth Macaw (also vulnerable to extinction, but less so than Lear's Macaw). It was described for science in 1856 by French naturalist Charles Bonaparte, nephew of Napoleon, and had been known from traded skins and some individuals in captivity, but there had always been doubt about its precise identification.

The name, Lear's Macaw, refers to Edward Lear, the nineteenth-century English author of limericks and nonsense verse and also a very talented artist. Lear painted live parrots at London Zoo for his work, *Illustrations of the Family of the Psittacidae, or Parrots* (1832), which included an example of the species which later was named after him. Lear painted 'his' macaw a couple of decades before it was described as a separate species by Bonaparte, and the captive specimen Lear painted was thought to be a Hyacinth Macaw and labelled as such in his book. The birds are similar, with gorgeous brilliant blue plumage, a yellow patch of skin near the bill and a yellow ring around the eye, but Lear's Macaw is noticeably smaller (approximately 75 cm/29 in. compared with 100 cm/39 in.).

A few years after its discovery in the wild the global population was estimated at only 60–80 birds, and so conservation efforts have been directed at protecting the species and its habitat. Today, the population is thought to number over 1,100 individuals and the species has been moved from 'Critically Endangered' to 'Endangered' – this is a conservation success story, though there is still some way to go.

The Lear's Macaw inhabits dry thorny areas where there are dense groves of licuri palms, and it feeds on their palm nuts, which make up around 95 per cent of its diet. This palm once covered an area of 250,000 sq. km (over 96,500 sq. miles) but has lost 98 per cent of its original range, so it is safe to assume that the species was once much commoner and more widespread, along with

OPPOSITE
When Edward Lear painted this bird from life in London Zoo in the late 1820s or early 1830s, he thought he was painting a Hyacinth Macaw (*Anodorhynchus hyacinthinus*). Only later did it become clear that this was a similar but different species. Although named after its painter, the bird remained a mystery until it was found in the wild 150 years later.

MACROCERCUS HYACINTHINUS.

Hyacinthine Maccaw.

E. Lear del. et lith. 3/4 Nat. Size Printed by C. Hullmandel

the trees. The remaining areas of the bird's preferred habitat are vulnerable to forest fires. New areas of licuri palms have now been planted to allow the Lear's Macaw to increase both its range and numbers.

In addition to decline caused by habitat loss, this beautiful species has been vulnerable to trapping for the bird trade, as Lear's Macaws can fetch very high prices. Better protection from this threat has helped the population increase. Another danger comes from farmers who occasionally kill the parrots if they eat their crops, particularly maize.

The macaws nest and roost in burrows excavated in sandstone cliffs. They first use their saliva to moisten the rock to soften it, and then with their beaks begin to excavate holes, which are further enlarged with their strong feet and claws. The breeding season coincides with Lent, but also, more importantly for the birds, with the peak output of licuri seeds.

ABOVE
Lear's Macaw is a smaller species than the Hyacinth Macaw. Its numbers in the wild remain very low but are thought to be increasing.

OPPOSITE
Lear's Macaw depends on the licuri palm for most of its food, and this specialization, coupled with huge losses of its required habitat, make it vulnerable to extinction.

There are now as many birds held in captivity as were once thought to make up the entire global population (74 individuals). In April 2015, São Paulo Zoo announced the first Lear's Macaw to be hatched in captivity anywhere in the world. The zoo holds a captive population of twelve adults, and this event raises hopes that a larger captive population could be established. Lear's Macaws are thought to live for up to 60 years in captivity, though this is largely a guess based on the lifespans of other parrots.

For most of the time since it was first described as a separate species, the Lear's Macaw has been something of a mystery bird. Once it was discovered in the wild it became a species of concern because of its small populations. Now, the future seems more optimistic for this beautiful species thanks to conservation efforts.

SPOON-BILLED SANDPIPER

Eurynorhynchus pygmeus

The Spoon-billed Sandpiper is a small wading bird that has an amazing spoon-shaped bill. In winter, it feeds in the soft muddy sediments on the coasts of Myanmar, Thailand and Bangladesh, moving its spatulate bill from side to side to sieve out small invertebrates. On its breeding grounds in Arctic Russia, 8,000 km (5,000 miles) to the north, it feeds mostly by pecking at small insects on or just below the surface of the water or the damp soil, and the highly unusual bill has no special advantage. Only by seeing the bird in its winter quarters does the unusual bill morphology really make sense – otherwise it looks like an evolutionary eccentricity.

In the 1970s the breeding population was thought to be in the order of 2,500 pairs, but 2011 estimates put the number at fewer than 100 pairs, and so the species has been listed as 'Critically Endangered'. At the present rate of decline the Spoon-billed Sandpiper could disappear in a decade. There may now be as few as 100 birds left in the wild and an ambitious conservation programme is under way to try to rescue it from extinction.

Spoon-billed Sandpipers return to nest in the Chukotsk and Kamchatka peninsulas in the Russian Far East bordering the Bering Sea in June as the snow melts, and then leave again in July. Studies on the nesting grounds suggest that breeding success and adult survival are within the bounds that would maintain the population, except that very few young birds return to nest at two years of age. The birds migrate through many countries, including North and South Korea, Japan and China as they travel to the wintering grounds. A known important staging area on their route are the massive mudflats of the Saemangeum Estuary of the Yellow Sea in South Korea, which have been partly reclaimed for agricultural and industrial land. This partial destruction of the intertidal area, completed in 2006, may have hastened the decline of the Spoon-billed Sandpiper, as an important link in the chain of stopover sites was severely damaged by this loss of estuarine feeding habitat.

ABOVE
As its name suggests, the Spoon-billed Sandpiper has a distinctive spoon-shaped bill. In summer it nests in Arctic Russia, migrating to the coasts of Myanmar, Bangladesh and Thailand to spend the winter. This illustration shows the bird with summer (left) and winter (right) plumage.

And once the Spoon-billed Sandpipers arrive in their wintering grounds they face trapping and hunting. Although the trapping, using fine nets, is mostly targeted at bigger species, such as ducks and larger waders, no birds that are caught are released; all are eaten. A study in Myanmar suggested that trapping rates were high: in the Bay of Martaban, the majority of subsistence hunters who were questioned knew of the Spoon-billed Sandpiper and probably caught individuals every year. It appears that the sites favoured as wintering areas by Spoon-billed Sandpipers are all ones with high levels of trapping, which might be enough on its own to drive the species' rapid decline. Since the young birds remain on the wintering grounds for most of their first two years they would be exposed to hunting pressure for all of that period – this may explain the low return rates of young birds.

One aspect of the international conservation programme is to attempt to provide hunters with alternative forms of food and to explain the plight of this special bird (easily identifiable because of its characteristic bill shape). In addition, a breeding programme is being established in the UK by the Wildfowl and Wetlands Trust at Slimbridge, where it is hoped to build up a captive population which could then be released to bolster what remains of the wild population when hunting pressure is lessened.

CALIFORNIA CONDOR

Gymnogyps californianus

The California Condor has the largest wingspan of any North American bird at 3 m (almost 10 ft). This magnificent carrion-eater soars on the thermals above the mountains and deserts of the southwest USA scanning the ground for dead cattle and deer, just as its ancestors, 10,000 years ago, searched for the bodies of mammoths and mastodons and for those of beached whales on the coasts. It is a survivor of the age when the North American continent supported herds of grazing herbivores roaming its plains and forests to rival those of East Africa.

Condors are one of the longest-lived birds, surviving to over 60 years. They first breed at six years and thereafter attempt to raise a single chick each year. As chick productivity is so low, relatively small reductions in survival will cause the population to fall.

The modern world of power lines, poisonous agricultural chemicals and habitat change is a difficult one for the California Condor. Ranchers, seeing the condors feeding on dead cattle, erroneously assumed that they had killed them and persecuted the birds accordingly. All these factors brought the California Condor population to a very reduced level, numbered in dozens of birds in the 1970s, and conservationists were afraid that after millennia of occupying the skies of North America the California Condor would become extinct.

After years of agonizing, the brave decision was made to take the entire wild population of California Condors into captivity and set up a captive-breeding scheme, which could then provide birds to reintroduce into areas still most suited to them. All California Condors then left in the wild were captured, the last one on Easter Sunday 1987, and San Diego and Los Angeles zoos became the main homes for this ancient bird.

ABOVE
Millions of dollars have been spent on captive-breeding the majestic California Condor to release birds back into the wild, and numbers are steadily increasing. This is one story where human intervention has had a positive effect on a threatened bird population.

OPPOSITE
California Condors have soared over the plains of North America for thousands of years searching for carrion. The mammoths on which they used to feed are now long gone, and the California Condor almost followed them to extinction.

Reintroductions began into California in 1991 and into Arizona near the Grand Canyon in 1996. As of 2014 there are over 400 California Condors on Earth, 225 in the wild and 214 in captivity. The same year also saw the first successful breeding in the wild in a new state, Utah, in Zion National Park, since reintroduction commenced. The bird's recovery has been aided by the banning of lead ammunition in much of its range, removing the very serious problem of lead poisoning from discarded deer carcasses left by hunters.

This is arguably the most expensive species conservation programme in US history, but it is, so far, a spectacular success. Today, a visitor to the Grand Canyon may see a condor circling high in the air, its enormous wings casting a shadow, just as they did over herds of mammoths in ancient days.

RIGHT
An illustration by Alexander Wilson of the California Condor, with its distinctive ruff of feathers at the base of its bald head and neck..

GREAT AUK

Pinguinus impennis

ABOVE
Found only in the North Atlantic, the Great Auk resembled the penguins of the southern hemisphere in morphology and ecology. It was a strong swimmer but could not fly.

A large flightless seabird of the North Atlantic, the Great Auk was once numbered in the millions but is now extinct. With its upright stance (standing around 80 cm or 30 in. tall), black-and-white plumage and short wings used for swimming underwater, it closely resembled the penguins of Antarctica. Its scientific name *Pinguinus* also alludes to penguins. In fact, it was probably the other way round, with penguins being named after the Great Auks, which may have got their original name from the Welsh 'Pen Gwyn', meaning 'white head' – although Great Auks had a mainly black head with a prominent white patch in front of the eye.

It is thought that Great Auks hunted co-operatively, feeding on a wide variety of marine fish, crabs and animal plankton in shallow, often coastal waters. They only came ashore to breed, nesting in colonies on remote, predator-free islands, laying a single egg in May on bare rock at up to 100 m (over 300 ft) from the shore. On land they walked slowly, holding out their wings for balance, and ran unsteadily, making them easy prey to predatory humans.

This is one case where we can be sure that extermination was a result of hunting pressure. The birds were a source of fresh meat for sailors, and fishermen also used Great Auk flesh to bait their hooks. But it was their thick down, which protected them from the cold Arctic seas and Arctic winds, that led to their extinction. The down was prized for making pillows.

Various laws were passed to protect the Great Auks from overexploitation, but they were too late and were almost impossible to enforce. As the bird became rarer and rarer, ironically, its eggs and skins became more and more highly prized by museums and collectors wishing to ensure that they had a specimen of a species that might go extinct – thus hastening that very fate.

RIGHT
Great Auks only came to land to breed and nested on isolated islands, giving them protection from most mammalian predators. They laid a single egg on the bare rock some distance from the shore.

OPPOSITE
On land, Great Auks were slow and relatively ungainly, making them easy prey for hungry sailors and fishermen. The Great Auk was hunted as a source of meat and as fishing bait, and its feathers and down were used for stuffing pillows.

OVERLEAF
Great Auks hunted for food underwater and are thought to have caught their fish co-operatively in flocks in the waters of the North Atlantic.

The last colony of Great Auks was in Iceland on 'Great Auk Island' (Geirfuglasker), where it was protected by high cliffs. But after volcanic activity the island sank and the birds moved to another nearby, which was accessible from one side. The final two birds here were caught and their egg smashed. This occurred on 3 July 1844, although a later record of a single bird off the Grand Banks of Newfoundland, Canada, is regarded as the last reliable sighting of this species.

The account of the last documented killing of a Great Auk, by Sigurður Ísleifsson, reads: '[I] caught it close to the edge – a precipice many fathoms deep ... I took him by the neck and he flapped his wings. He made no cry. I strangled him.'

147

Alca impennis

Revered & Adored

Raven • Sacred Ibis • Goldfinch • Andean Condor Albatrosses • Owls • Eagles • Red-crowned Crane Resplendent Quetzal • Hoopoe

ABOVE Golden Eagle.

We are captivated by birds for many reasons, but there are some whose relationship with humankind far surpasses their simple biological characteristics or their utility to us. Their meaning transcends the merely physical and passes into what they symbolize to us rather than what they are. Sometimes this symbolic importance is rooted in the bird's biology; often it is not, or if so, only very loosely. But that is not to say that these meanings are any less real. On the contrary, for most of human existence they have perhaps been the predominant types of relationship we have had with birds.

Wherever humankind has travelled throughout the world and in all ecosystems we have encountered birds. Some became familiar parts of our everyday lives, while others were rarer and stranger. Many species have been thought to have great mystical significance for both good and evil, predicting our fates or tied up with our destinies. Some have been closely associated with religion in one way or another, such as the Goldfinch and the Hoopoe, and were seen as the symbols or even manifestations of gods, for instance the Sacred Ibis.

These days we tend to steer clear of anthropomorphism when discussing the natural world, but we still use descriptions derived from birds and are happy to give ourselves avian characteristics, such as 'cocky', 'hen-pecked', 'lovey-dovey', 'hawk-eyed'. And conversely, we also attribute human qualities to some birds – the wise owl, the trickster Raven and the majestic eagle, king of the birds. Our relationship with birds is clearly not one of equals, but it seems that we treat birds as being more human-like than many other animals, even many mammals to which we are more closely biologically related, but perhaps, in our minds, not as emotionally connected.

Of course, particular birds have become important to specific people, depending on where they live. They can be closely interwoven with cultural identity, for instance the Andean Condor, the Red-crowned Crane and the Resplendent Quetzal, and can also carry different meanings for different cultures, such as the albatross. Here we can examine just a very few cases, from across the world, where birds have become closely entwined with human cultures in myth and legend, in religion and ideology, in art and literature, and as national symbols.

RAVEN

Corvus corax

Its large size (up to 67 cm/24 in.), glossy black plumage and deep guttural calls make the Raven an easy species to see and identify. It is also quite widespread, being found across the whole of North America, Europe and most of Asia. To a large extent it feeds on carrion, and is big enough to displace most other birds from a carcass, which is possibly why the Raven is often associated with death. Its acrobatic display flights and loud calls may have led also to the belief that it is a trickster and playful species.

ABOVE
This nineteenth-century Japanese painting captures the Raven's bulk and power. The world's largest perching bird, the Raven is found widely across a range of habitats in the northern hemisphere.

OPPOSITE
In some legends, the Raven started with white plumage but became black when it stole fire for humans to use.

Mythology and symbolism involving the Raven can be found in many cultures, but perhaps especially so for the peoples of the Pacific Northwest Coast of North America. Here the Raven is variously credited with discovering humankind, introducing men and women to each other for the first time, creating the world, stealing fire for humans to use and fixing the sun, moon and stars in the sky. In many of these tales the Raven combines his role of provider to humankind with his ability to trick other animals and birds. For example, the gull was said to have been given the sun, moon and stars in a box, but refused to open it to allow them out, and so the world was in constant night. It was the Raven who, after failing to persuade the gull to open the box, stuck a thorn in the bird's foot and pushed hard until the gull dropped the box and the world received light.

Another story has the Raven falling in love with the daughter of the eagle. And when the Raven saw that light was kept hanging on the sides of the eagles' lodge, he waited for his chance and stole the sun, moon and stars and hung them in the sky. In some legends, the Raven was originally a pure white bird, but acquired its black plumage either from the smoke from the embers of the fire it stole to give to humankind, or because it was flung on a fire – where its

2
1

RIGHT
A shaman's rattle, from the Pacific Northwest Coast of North America, in the shape of a Raven, featuring a figure on its back with legs akimbo. The Raven appears in many legends and is usually portrayed as a cunning and mischievous character.

BELOW
Ravens lay between four and six pale eggs in nests of sticks, which are often on cliffs but are sometimes found in trees. They are early nesters, with eggs commonly laid in late February.

feathers were forever singed – by human hunters punishing it for warning the buffalo of their approach. The importance of the Raven in these cultures of the Northwest Coast is reflected in its frequent representation in their art, in the form of masks, rattles, canoe prows and elements in totem poles.

Elsewhere, the Raven has played a variety of roles. In Norse myth, Odin is accompanied by two Ravens (Huginn, 'thought', and Muninn, 'memory') who bring back news of the land of men to the god from their flights around the world. In Judaic tradition, the Raven taught Adam and Eve how to bury Abel after he was slain by Cain, and the Raven was also the first bird sent out from the Ark by Noah to search for signs of dry land. In the New Testament, Christ exhorts us to put our trust in God: 'Consider the ravens, that they sow not, neither reap; which have no store-chamber nor barn; and God feedeth them: of how much more value are ye than the birds!'

In Britain tame Ravens live in the Tower of London, and there is a legend that as long as the Ravens remain, the country will be safe. But rather than being an ancient legend, it is now thought that this is a relatively modern Victorian invention (an earlier Welsh legend has it that the head of the Celtic giant King Bran was buried at the location of the White Tower, facing France, also to protect the country). Ancient or recent, this superstition is firmly fixed in modern mythology.

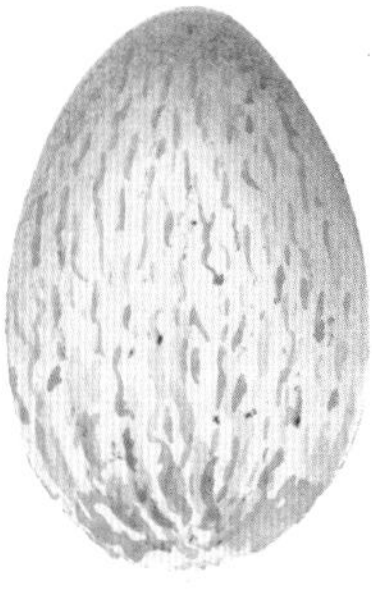

The Raven is the world's largest passerine (perching bird) and, perhaps befitting its stature, this all-black bird, with beady eye, big bill, croaking call and questionable dietary habits has found a special place in our imagination, surpassing that of many of its prettier, more colourful and more tuneful relatives.

SACRED IBIS

Threskiornis aethiopicus

ABOVE
A long-legged wetland bird, the Sacred Ibis feeds on a wide range of small vertebrates including fish, frogs and birds, as well as many large invertebrates.

BELOW
Sacred Ibises nest in colonies in trees. Their large nests are made from twigs and sticks, and the normal clutch size is two or three eggs.

In ancient Egypt the Sacred Ibis, a wetland bird, was regarded as one of the manifestations of the god Thoth, who was sometimes portrayed as a baboon but more often as a man with the head of a Sacred Ibis. Thoth was the god of wisdom, writing and science and also played a role as mediator in maintaining order in the universe, ensuring that neither good nor evil, order nor chaos, held complete dominion. It was Thoth, in fact, who was said to have invented writing, and the Sacred Ibis is depicted in the form of a hieroglyph.

An important element of the worship of Thoth was the offering of mummified ibises. Over four million bodies of Sacred Ibis were found in the catacombs of Tuna el-Gebel, and 1.75 million at Saqqara (both sites in Middle Egypt, close to the Nile). Birds were bred in captivity and reared for this cultic and devotional purpose, as well as being caught in the wild. Sanctuaries were established across Egypt for the purpose of breeding Sacred Ibises, and land was set aside to grow food to feed such large numbers of captive birds.

The birds were killed by having their necks broken and were then prepared for mummification. Often items such as snails were put into the birds' beaks as food for the afterlife, and the mummified birds were wrapped in linen and placed in pots. This great enterprise was supported by royal subsidies, but perhaps also by a tourism industry of pilgrims. The scale of the undertaking was immense and

RIGHT
The mummified bodies of Sacred Ibises were brought back from Egypt by scholars, including Vivant Denon on Napoleon's Egyptian expedition. These remains were used by naturalists in their study of theories about evolution.

OPPOSITE
Appropriately for a bird associated with the ancient Egyptian god of writing, the Sacred Ibis is the symbol of the British Ornithologists' Union. The organization's international journal, *Ibis*, was first published in 1859, at a time when the Sacred Ibis still lived in Egypt.

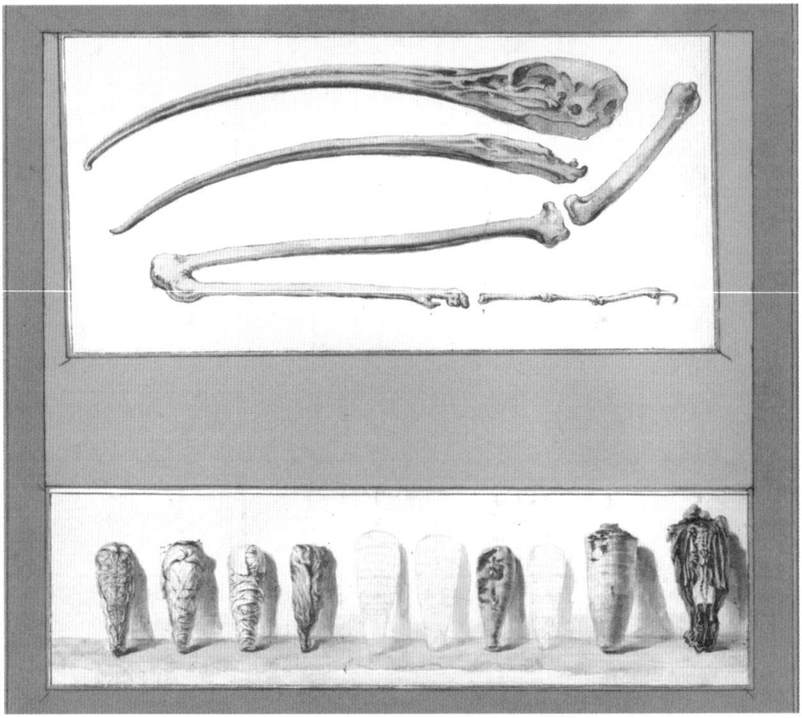

bears testimony to the importance of the bird, and by extension of Thoth, to the ancient Egyptians. Sacred Ibises were also common in the form of amulets and statues, and were represented in tomb paintings.

In the nineteenth century the long-mummified bodies of Sacred Ibises were compared with their living equivalents by French naturalists such as Georges Cuvier. The lack of discernible differences in structure, over a period of about 3,000 years (then believed to be a large proportion of the Earth's existence) was taken to be important evidence for a lack of evolution in species. Cuvier summed up knowledge about the birds, borrowing from the writings of Herodotus and Pliny: 'Every one has heard of the Ibis, the bird to which the ancient Egyptians paid religious worship ... which they allowed to stray unharmed through their cities, and whose murderer, even though involuntary, was punished by death; which they embalmed with as much care as their own parents.'

In ancient Egypt, the Sacred Ibis, with its distinctive white plumage, black neck and head, and curved beak, would have been a familiar sight, but desertification and habitat loss mean that it is no longer found there, and has not been seen in any numbers for over a century. There are thoughts of bringing the Sacred Ibis back to Egypt through a reintroduction project, but this would entail first ensuring sufficient areas of suitable safe habitat.

THE IBIS,

A

QUARTERLY JOURNAL OF ORNITHOLOGY.

EDITED BY

ALFRED NEWTON, M.A.,
PROFESSOR OF ZOOLOGY AND COMPARATIVE ANATOMY
IN THE UNIVERSITY OF CAMBRIDGE,
F.L.S., V.P.Z.S., ETC., ETC.

VOL. V. 1869.

NEW SERIES.

Ibidis interea tu quoque nomen habe!
—Ovid.

LONDON:
JOHN VAN VOORST, 1 PATERNOSTER ROW.
1869.

ABOVE
This mid-seventeenth-century trompe-l'oeil by Carel Fabritius depicts a common scene at the time: a captive Goldfinch – prized for its attractive plumage and song – chained to a perch on the wall.

OPPOSITE
In the wild, Goldfinches lay their first clutches in May and may successfully rear up to three broods each year. Five eggs are laid in a nest located in a fork between branches. The young hatch after two weeks and leave the nest just over two weeks later.

GOLDFINCH

Carduelis carduelis

The European Goldfinch is a common farmland species across much of Europe and western Asia; it is a familiar bird, with its bright red face and striking golden-yellow wing-bars. It is also a pleasing songster, and the combination of bright plumage and liquid, twittering song made it a favoured cage bird kept in captivity (as famously depicted in the painting by Carel Fabritius), which would delight both the eye and the ear. In the 1860s European colonists took Goldfinches with them to New Zealand and introduced them there to remind them of home. These days it is a bird increasingly seen in gardens, even urban ones, as it is attracted to seed feeders.

The collective noun for a group of finches, but especially Goldfinches, is a 'charm', which seems fitting for such a dainty and well-regarded bird. All these feelings of affection for the Goldfinch were expressed by the nineteenth-century English nature poet John Clare, using a common English name for it, the redcap:

The redcap is a painted bird
and beautiful its feathers are;
In early spring its voice is heard
While searching thistles brown and bare;
It makes a nest of mosses grey
And lines it round with thistle-down;
Five small pale spotted eggs they lay
In places never far from town

Much earlier, Geoffrey Chaucer had described the cook in the *Canterbury Tales* as being as 'merry as the goldfinch in the woods'.

OPPOSITE
In his painting *Madonna and Child* (1480), Carlo Crivelli shows a Goldfinch clasped tightly to the Christ Child's breast. The bird, and the cucumber above left, are symbols of goodness and hope, while the apples and fly symbolize evil and sin.

BELOW
Marginal illustration of a Goldfinch from the *Sherborne Missal* – one of many birds depicted in this early fifteenth-century book.

But alongside, or rather at a deeper level than, its attractiveness as a familiar bird, the Goldfinch held a profound symbolic meaning in a religious context for our forebears. Goldfinches feed on seeds, including teasels and thistles, and these spiny plants had associations not only with healing, but also with Christ's crown of thorns when on the cross. The golden wing-bars of the Goldfinch are thought to be evocative of healing, and the blood-red feathers of its face (in a similar way to the Robin's red breast) were said to have gained their colour when the birds tried to remove Christ's crown of thorns. Thus the Goldfinch was closely identified with the Passion of Christ and with redemption and healing. It is for this reason that the European Goldfinch appears in a huge body of medieval and Renaissance paintings, often ones of religious significance.

In Carlo Crivelli's *Madonna and Child* (1480) the Christ Child is clutching a Goldfinch tight to his chest, and in Raphael's *Madonna of the Goldfinch* (1505–6) the child John the Baptist is offering the Christ Child a Goldfinch. There are many more examples of such works, and in them the Goldfinch is a symbol of good rather than evil, and of hope rather than despair. It seems clear that the Goldfinch was a potent symbol in medieval times and that its frequent appearances in art would have been understood by the populace in a way that we modern-day observers of those paintings may not fully appreciate.

But as well as these sacred and lofty allusions to redemption, the Goldfinch has more earthy associations, being linked with human sexual activities, which is probably why a giant Goldfinch is depicted in Hieronymus Bosch's painting *Garden of Earthly Delights* (*c.* 1510). In our own more secular and prosaic age, the Goldfinch is perhaps simply a pretty bird with an attractive voice. It brings vivid splashes of bright colour into our lives that many other small and familiar European birds lack. In cold northern climes, the Goldfinch also brings a touch of the exotic.

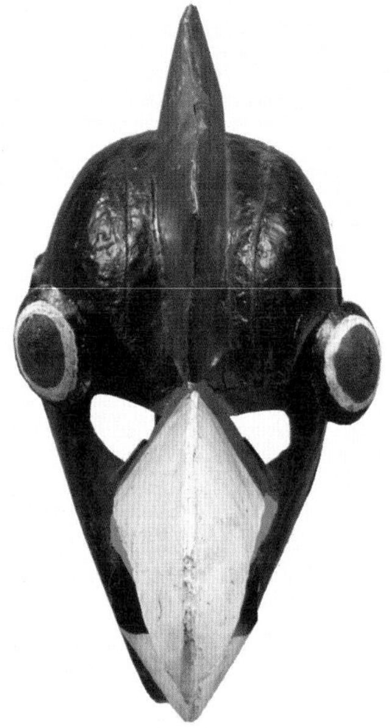

ANDEAN CONDOR

Vultur gryphus

ABOVE
An Andean Condor mask, made of plaster and glass, from Bolivia.

OPPOSITE
Andean Condors in their natural habitat. The birds in the foreground are an adult (left) and a juvenile (right). The flying bird illustrates the broad wings on which the Andean Condor soars high above the mountainsides and valleys, searching for carrion.

For thousands of years the human inhabitants of the Andes looked up into the skies and saw the Andean Condor, circling above them on its 3-m (10-ft) wings, and thought of their gods looking down on them from on high. The Inca believed that the Andean Condor brought the sun into the sky every morning and represented the gods of the 'higher plane', such as Inti the sun god, their most important god, from whom Andean rulers were believed to be descended. The plane of men was represented by a puma and the underworld by a snake.

And as the Andean Condor looked down, it was searching for carrion, mostly the bodies of large animals such as llamas, alpacas, deer, rheas and guanacos, though now horses and cattle are added to the range of species they feed on. Although Andean Condors are carrion-feeders, some ranchers believe that they actively kill their stock, and this has led to cases of deliberate poisoning and other forms of harm. Numbers of Andean Condors have fallen, and persecution of this bird, which is both long-lived (individuals can reach up to 70 years in captivity) and slowly reproducing (it first breeds at 5–6 years), will have played a part in this, but habitat loss and accidental poisoning seem to be the main factors driving the decline.

Andean Condors have mostly black plumage, with white wing feathers and a white ruff around the neck, which is mostly bald. Adult males also have a fleshy comb on the top of their heads. The bald head and neck of these birds are an adaptation for carrion feeding, for which it is also well equipped with a strong hooked beak.

This large bird of the high Andes is the national symbol of Argentina, Bolivia, Chile, Colombia and Ecuador. These constitute most of the countries in

VULTUR GRYPHUS (ECUADOR)
ad. (*left-hand fig.*) and juv. (*right-hand fig.*)

ABOVE
On a prairie in the Andes, an Andean Condor – with others in the air above – is disturbed from the carcass of a horse by riders attempting to lasso it.

which it occurs, with the exception of Venezuela, where it is very rare, and Peru (which chose the Andean Cock-of-the-Rock). It has been depicted in Andean art in ceramics and textiles for at least 4,500 years, and is regarded as a symbol of power and a messenger of the gods.

In the last few centuries, following the arrival of European invaders in South America, the Andean Condor has been the central feature of a now-contentious ritual in parts of Peru. In the Yawar festivals a wild Condor is captured, adorned with textiles and led through the village streets by its wingtips. It is then strapped to the back of a bull and the pair engage in a bullfight. The symbolism is clear: the bull represents the Spanish invaders and is killed by the men of the village aided by the Condor, which represents the Andean people riding on the bull's back, scratching with its claws.

In this ritual, which occurs in at least 55 villages in Peru (where the total Andean Condor population is only around 400 birds), the Condor is the hero and is feted. But it is a rough ritual and many of the participating birds are injured; 10–20 per cent are killed by accidental injuries from the antics of the rampaging bull, even though the aim of the event is to release the Condor in triumph. If the Condor is killed or injured it signifies bad luck for the village, but of course it is worse luck for the bird itself. Villages are fined if the Condor dies and it casts a shadow over the whole celebration of the symbolic victory of the local people over the European oppressor.

For some, nothing could sum up better the complexity of our relationship with birds and wildlife and the environment as a whole than the evolving relationship we have with the Andean Condor. For thousands of years it has been revered, but its population is now declining thanks to a combination of ill-informed beliefs about its impacts on livestock and unintentional pressures from a growing human population. At the same time this iconic bird has been incorporated into a more modern ritual, in which it is the hero in a political allegory, but in which it is often the actual, if not the symbolic, loser.

ALBATROSSES

Diomedeidae

ABOVE
An engraving by Gustave Doré from his illustrated edition of Samuel Taylor Coleridge's *Rime of the Ancient Mariner* (1877). This illustration shows the albatross befriending the sailors, who do not look at all delighted to see the bird.

These magnificent seabirds have the largest wingspans of any bird – up to 4 m (12 ft) in the case of the Wandering Albatross (*Diomedea exulans*). A few of the 22 species of albatross are found in the northern hemisphere, but most circle the Southern Ocean on their stiff wings, braving the storms and skimming the waves apparently almost effortlessly. They are long-lived birds – up to 60 years – and spend the majority of their time at sea, travelling huge distances and only coming to land to breed, mostly on isolated islands. A Laysan Albatross (*Phoebastria immutabilis*) lived for at least 63 years, nesting on Midway Atoll, Hawaii, and probably raising at least 35 chicks.

Albatrosses have entered the mythology and culture of many peoples. For instance, in Hawaii the albatross is associated with the god Kane. These birds have long been highly regarded by the Māori people of New Zealand, who ate them, used their bones for fish hooks, spear tips, ornaments, flutes (especially the long wing bones) and even chisels for tattooing, and their feathers for decoration and clothing. Albatross feathers, *titapu*, were worn by those of high rank and also decorated war canoes when they put to sea.

A particular Māori decorative pattern, known as 'Tears of the Albatross', *roimata toroa*, is woven into mats and used on wall panels in meeting houses. The albatross was reputed to shed tears when mistreated by man or missing its distant marine home. The idea that the albatross cries tears probably derives from the ability of the bird to drink sea water and get rid of salt by secreting a concentrated saline solution from its salt glands and through the nostrils. The legend is thus based on sound biological observation rather than whimsical imagination.

Western cultures only encountered albatrosses in the eighteenth century. Sailors observed these impressive birds confidently gliding above the waves even

RIGHT
An albatross bill is long and strong, with a large hook at the end to grab hold of prey, including fish and squid. Albatrosses can drink seawater and secrete excess salt from their salt glands and through their nostrils.

OPPOSITE
The Wandering Albatross has the longest wingspan of any bird, measuring up to 4 m (12 ft).

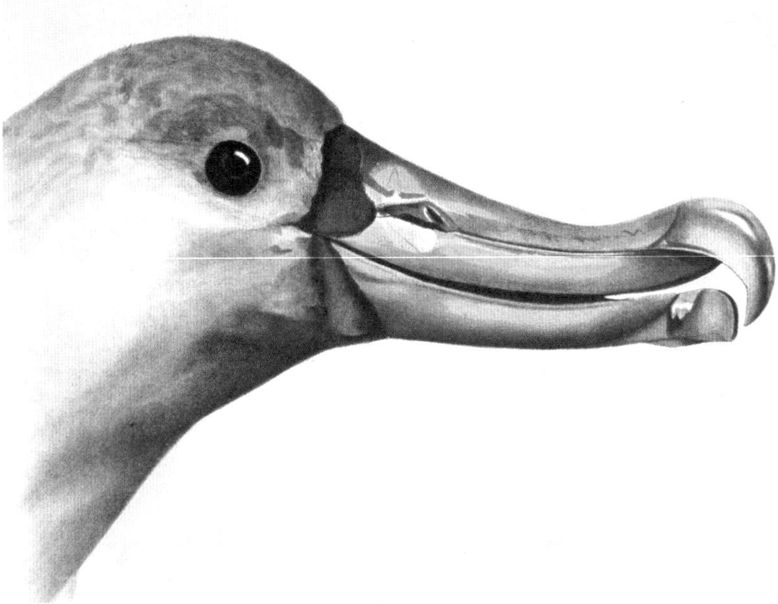

in storms that terrified those on board ships, and thought that the albatrosses might be the embodiment of dead mariners sent to watch over them. Similar stories had attached to gulls and dolphins before albatrosses were known.

For many, Samuel Taylor Coleridge's *The Rime of the Ancient Mariner* (1798) is their main connection with this bird of the Southern Ocean. The long poem tells the story of a sailor who shot an albatross with a crossbow, even though the bird had guided his ship out of pack-ice. His fateful act brings bad luck to his ship and its crew, and they hang the dead albatross around the sailor's neck to remind him of his evil deed. He is redeemed when he comes to love and appreciate the wildlife of the oceans, and finally the albatross drops from his neck.

Too often the term 'an albatross around one's neck' is used to refer to a burden, but the true message of the poem is that we should respect all nature. More than two centuries after the poem was published, the world's albatrosses are threatened with extinction, mostly as a result of our industrial-scale fishing techniques. Previously they were caught in huge drift nets, and now they are killed in large numbers by longline fisheries as they take baited hooks. A further threat comes from the accumulation of plastic detritus in the oceans, which the birds inadvertently swallow. It causes a particular risk for chicks, which choke on it when food is regurgitated for them. Despite the albatross's mythical status, and although we have stopped shooting them with crossbows, it seems we have not yet fully learned the lesson of the Ancient Mariner.

PLATE 40

WANDERING ALBATROSS

[One-fifth natural size]

OWLS

Strigiformes

Owls have had a mixed press over the millennia. As mostly solitary, nocturnal creatures, with soft plumage and silent flight, they have often been regarded as birds of ill omen, stealthy messengers who come at night to warn or chastise. The nocturnal hoots and screeches of owls are thought supernatural and have not impressed poets or writers, being difficult to interpret as lullabies or serenades. The cries of owls punctuate the night and disturb sleepless humans right around the world, and generally speaking they have been seen as bringers of bad news and portents of misfortune.

There are about 200 species of owl found across the globe, except in icy wastes (although the Snowy Owl feeds on lemmings far north of the Arctic Circle). They hunt using their forward-facing eyes and keen hearing. Owls' disc-shaped faces concentrate and focus the sounds into the ears, which are located asymmetrically. The asymmetry is thought to aid the precision of detection of the sources of sounds and thus contributes to successful hunting. The basic owl form is of a fairly standard shape, but the sizes vary from the fish-catching Blakiston's Fish-Owl (4.5 kg or 10 lb) of Russia to the Elf Owl (31 g or 1 oz).

For the Kikuyu of Kenya, the Aztecs and Maya of Central America and the Cherokee in North America, owls were thought to be harbingers of death. In Malaya and Arabia owls were thought to kill and carry off newborn babies. In such circumstances owls were often themselves victims, being killed and sometimes chased at night with brands of fire and, if caught, set alight to put an end to any perceived bad luck. In many parts of the world, people resorted to a range of different actions to ward off the misfortunes brought by owls, including nailing a dead owl to a barn door (widespread in Europe), throwing salt in the fire (France) and shouting 'Salt and pepper for your mammy' (Jamaica).

ABOVE
There are around 200 species of owl, which largely conform to a pattern of forward-facing eyes, facial discs and sometimes tufts of feathers resembling ears (although the real ears are set in the facial disc). Most owls are largely or entirely nocturnal, and many hunt primarily by hearing.

OPPOSITE
Barn Owls (*Tyto alba*) are found on every continent on Earth except Antarctica. Owls have sometimes been seen as harbingers of death or misfortune, but they are also associated with wisdom.

ABOVE
A silver tetradrachm coin depicting the owl of Athena (*c.* 480–420 BC) or Little Owl. This species can still be seen today on the Acropolis in Athens, where Athena's temple, the Parthenon, is situated.

OPPOSITE
A Little Owl painted by a Dutch artist, Cornelius Nozeman, around eighteen centuries after the Athenian coin (illustrated above) was made. The Little Owl is found across much of Europe and central Asia.

Sometimes a culture's attitude towards the owl can be more ambiguous. In China owls are found in Neolithic art of the sixth millennium BC onwards and feature especially in bronze objects from the Shang dynasty from the early second millennium BC. It has been suggested that the owl had an important role in Shang beliefs, and was perhaps a totemic animal and thought to be a messenger between the human and spirit worlds.

In Hinduism, each deity has a *vahana* (an animal that carries the god or goddess). Lakshmi, the beautiful goddess of plenty and wealth, chose an owl as hers. The owl is seen as a creature which prowls at night because it goes blind in the day, and this symbolizes humankind's tendency to be attracted by secular rather than spiritual wealth. By choosing the owl as her *vahana*, Lakshmi is teaching us to open our eyes to the wisdom inside us all.

In more recent times, from Classical Greece to Harry Potter, owls have been seen in a more positive light as a source of wisdom. The Greek goddess Athena had the owl, almost always depicted as a Little Owl (*Athene noctua*), as her sacred bird, and so the owl became associated with philosophy and wisdom. Was it perhaps because the Little Owl is one of the least nocturnal and smallest of owls that it was more easily associated with a beneficial attribute such as wisdom? Little Owls inhabited the Acropolis and were portrayed on Athenian coins. Athena herself was often portrayed with a Little Owl sitting on her hand. We now tend to see the owls' ability to fly by night as demonstrating intelligence and perspicacity, and talk about 'wise old owls'.

Numerous superstitions have attached to owls see them in different lights. They are associated with witchcraft and magic, but also with medicinal powers. In England in the past various remedies could be obtained by eating owl eggs or owl broth, including a lifetime cure for alcoholism, better eyesight and relief from whooping cough.

Throughout the ages and around the world, owls have been a source of mystery and even fear for us, largely because they inhabit the night when we feel at our most vulnerable and afraid. To many they have seemed like spirits travelling the darkness, and in our imaginations we perceive them as evil and harmful to us. But we also have an attitude of respect and even admiration for them, and their useful ability to catch and thereby control vermin is recognized. Perhaps we have acquired some of the proverbial wisdom of the owl.

EAGLES

Aquilae

Eagles are large, majestic birds of prey, mostly with strong large beaks for tearing flesh and powerful talons for subduing and killing prey such as small deer, hares and birds up to the size of Wild Turkeys. However, the species given this name are not all closely related to each other genetically – in fact the taxonomy of eagles is in a state of flux. And there are a few small eagles too, such as the Little Eagle of Australia, which is no larger than a falcon.

The eagle is commonly known as the king of the birds (despite the claims of the Goldcrest, see p. 26) because, with its size, power and soaring flight, it appears to command the earth from above. Eagles also seem to have freedom – to go where they like unmolested and without fear. That vision may appeal to many a nation, and eight countries – from the USA (Bald Eagle) to Panama (Harpy Eagle; p. 58), the Philippines (Philippine Eagle), Scotland (Golden Eagle), Malawi, Namibia, Zambia and Zimbabwe (all African Fish-eagle) – have an eagle as their national bird. The Bald Eagle became the USA's national bird in 1782, chosen for its long life, great strength and majestic looks.

In Greek mythology the eagle was closely associated with Zeus, the king of the gods; the bird was his messenger and he transformed himself into an eagle when he abducted Ganymede to serve as his cup-bearer on Olympus. In Classical history, the eagle has had military significance too. The Roman legions each had an 'eagle', carried by a standard-bearer called an *aquilifer*. For a legion to lose its eagle in battle to an enemy was a matter of great shame, and one that the Romans sought to correct as quickly as possible. Many later empires also chose the eagle as their heraldic symbol.

The Native American warbonnets of the Plains Indians were made from the tail feathers of Golden Eagles. Sometimes the feathers were obtained by hiding under a dead animal for days until an eagle landed to feed on the carcass and then grabbing it by the legs, but at other times young eagles were taken from nests, reared in captivity and had their tail feathers plucked several times. Eagles

OPPOSITE
A Tawny Eagle (*Aquila rapax*) by Louis Agassiz Fuertes, showing the powerful beak used for tearing skin and pulling at flesh.

quila-rapax raptor
N'Jabarra, Gojam
-Mar. 23. 1927-
-(living bird)-
Tawny Eagle

GOLDEN EAGLE, FEMALE.

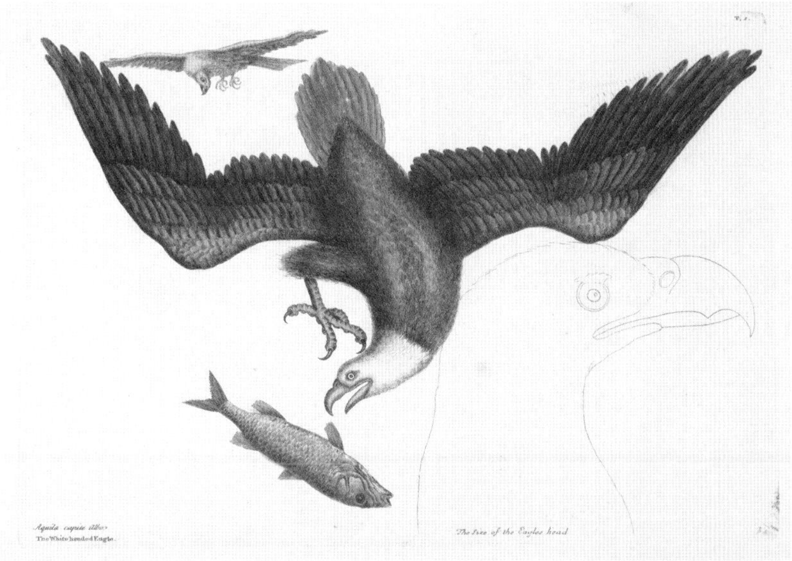

had a special significance for many Native American peoples and feature prominently in mythology and ceremonies. The Hopi of the American Southwest have a particular reverence for and relationship with the eagle and perform the Eagle Dance, as do other peoples in the same region.

In Hinduism, Vishnu's *vahana* (an animal that carries the god or goddess) is an eagle (or possibly a kite) called Garuda. When Vishnu travels, sometimes accompanied by Lakshmi, they are carried by Garuda. Garuda was said to be the sworn enemy of the serpent race – perhaps a reference to the Short-toed Eagle that does eat snakes.

It may seem strange to find this powerfully predatory bird depicted in a Christian church, but lecterns often take the form of an eagle. In Christian tradition, the eagle was thought to be able to stare unflinchingly into the sun, as Christians could stare unflinchingly at the truth of the Bible. But the eagle was also a symbol of Christ and the emblem of St John the Evangelist.

The eagle as a symbol has also entered contemporary culture. From American football's Philadelphia Eagles to Nigeria's national soccer team the Super Eagles, many teams want to be associated with the eagle, presumably hoping to absorb some of the bird's strength, speed and dominance.

Eagles are almost always associated with power, and often military power. The fact that they are among the largest birds of prey, can fly fast, high and far, and are formidable hunters, all combine to create an image of this bird that throughout history we have wanted to incorporate into our myths, religions, armies, national identity, and now even our businesses and sports teams.

ABOVE
A Bald Eagle (*Haliaeetus leucocephalus*) by eighteenth-century artist Mark Catesby. The bird looks as though it may have dropped a fish, which it would seek to recapture with its talons.

OPPOSITE
A Golden Eagle (*Aquila chrysaetos*) subdues a rodent. Bronze eagles are often depicted in a similar stance in lecterns in Christian churches, with their unflinching gaze directed at the congregation. There are numerous biblical references to the eagle, which is admired for its strength and speed.

ABOVE
The Red-crowned Crane features frequently in Japanese art, as in this woodblock print by Hokusai. The affection for Red-crowned Cranes in Japan has helped to stabilize their numbers in areas where birds are fed with grain, a sight that in turn attracts crowds of tourists. However, overall the future of this bird remains uncertain.

RED-CROWNED CRANE

Grus japonensis

The sight of Red-crowned Cranes dancing in pairs in the late winter, often against a background of snow, as they re-establish their pair bonds for the breeding season can appear almost ethereal – white birds dancing in a white landscape. The two members of a pair seem only to have eyes for each other as they jump into the air, spread their wings and fold and then stretch their necks. They weave around each other and call as though in ecstasy.

While the plumage of these large cranes (150 cm or almost 5 ft tall) is predominantly white, they also have a black face and neck and wing-tips. A distinguishing feature is of course the patch of bright red featherless skin on the top of their heads. Unfortunately, the Red-crowned Crane is also endangered and is one of the rarest cranes in the world, with fewer than 3,000 left in the wild. There is a resident population in Japan and a migratory one that breeds in Russia, China and Mongolia and winters in China, South Korea and North Korea.

This beautiful and graceful bird appears frequently in the arts of China and Japan, and has great symbolic meanings. Cranes were depicted in Chinese tombs over 3,000 years ago, and in art immortals are often shown riding cranes, themselves symbols of longevity or immortality, perhaps because they are long-lived birds. In Japan, cranes are said to live for 1,000 years. A thousand origami cranes, held together by string, is a traditional wedding gift from a father to the newlyweds wishing them a thousand years of happiness and prosperity. Since it mates for life, the crane is also a symbol of fidelity.

RESPLENDENT QUETZAL

Pharomachrus moccino

ABOVE
Of great cultural significance to both the Maya and Aztec civilizations, Resplendent Quetzals have long been protected from harm by law. The tail plumes of this bird were once used as currency.

This aptly named bird has been of great cultural significance in Central America, particularly Guatemala, for over 2,000 years. Associated with the feathered snake god of the Maya, Kukulkan, and the Aztec version, Quetzalcoatl, the bird symbolizes freedom and is depicted on numerous ancient artifacts as well as being almost ubiquitous on modern trinkets and textiles.

The Resplendent Quetzal is a species of trogon, a group of 39 tropical species found in Central and South America, Africa and Southeast Asia. The females are a beautiful lime green, and the males also have spectacular red breasts and bright white tails; even so, they might not warrant the name 'resplendent' were it not for the four elongated iridescent green plumes of up to a metre (3 ft) in length that stream and flutter behind the tail of males and triple the overall length of the bird. These feathers were very highly prized and traded in large numbers over great distances – Aztec wars were even fought to secure supplies of them. The feathers were used in ceremonial head-dresses and one spectacular example, incorporating 450 quetzal plumes and perhaps originally belonging to the Aztec ruler Moctezuma (Montezuma) II, is displayed in the Vienna Museum of Ethnology.

The Resplendent Quetzal has only recently been bred successfully in captivity (this is one reason why it is seen as a symbol of freedom – reputedly it cannot tolerate the loss of its liberty) and thus wild birds were the only source of the precious feathers. Quetzals were caught on sticks covered with bird lime to which the birds became stuck, having been lured by food scattered as bait. But the trapped birds were released once their tail plumes had been plucked from them. Killing a Resplendent Quetzal was regarded as taboo and was punishable by death.

ABOVE
Resplendent Quetzal feathers adorning headdresses and staffs, from the Codex Mendoza of Mexico, sixteenth century. Several illustrations feature such feathers, which are associated with people of high social rank.

OPPOSITE
This early nineteenth-century illustration by John Gould shows the Resplendent Quetzal's long, iridescent tail plumes to great effect. The plumes are around twice its body length, and float and undulate behind the bird as it flies.

Legend has it that when the Maya prince Tecun Uman lay dying at the hands of a Spanish conquistador, a male Resplendent Quetzal flew down and pressed its breast against the bloody chest of its countryman and thus acquired its red feathers. A further legend states that Resplendent Quetzals sang beautifully before the Spanish Conquest and will regain their voices when their forests are truly free again.

Although under threat because of deforestation, the Resplendent Quetzal is still a reasonably familiar sight through its range, and so the visitor to Central America can sometimes enjoy the shimmering colours of the bird, which inhabits rainforests and feeds on wild avocados. As a male flies, rather weakly, through the forest or across a road, its bright colours and streaming plumes can still delight the senses. And so it seems fitting that a bird which has been valued so highly for so long, now gives its name to the currency of Guatemala, and is depicted on the quetzal coins and banknotes.

HOOPOE

Upupa epops

A widespread bird of Europe, Africa and Asia, the Hoopoe is a visually striking one too. Its pinkish-buff body plumage is highlighted by conspicuous black-and-white stripes on its broad wings and its tail. It also has a long, slightly curved beak, which it uses to probe the ground and the bark of trees for insects and grubs. But perhaps its most distinctive feature is the long crest at the back of the head which is raised, like a Mohican hairstyle, if the bird is excited. In different regions and languages the bird often derives its name from its song 'Hoo-poo' or 'Hoo-poo-poo'. All in all, the Hoopoe is a bird that attracts attention for both its appearance and its voice.

ABOVE
A bird of open woodland and dry grasslands, the Hoopoe has a far-carrying three-note song, recalling the cuckoo in tone. Both its song and its plumage proclaim its existence.

OPPOSITE
The striking black-and-white striped plumage of the Hoopoe is conspicuous as the bird stalks across the ground, but even more so in flight. The crest can be raised and lowered according to mood and for display.

Perhaps because of this, Hoopoes were depicted in the art of ancient Egypt (where they were also used in cures) and Minoan Crete. They also have a special place in the Jewish and Islamic traditions. One story tells how King Solomon was walking through the desert and was in danger of being overcome by the heat when he was saved by Hoopoes flying over him to provide him with cool shade. Solomon offered the birds a reward, and after much discussion they asked for a crown of gold. Solomon warned them they would regret it, but granted the wish. Once people discovered that Hoopoes were crowned with gold, they started trapping and shooting arrows at them, and the Hoopoes came back to Solomon admitting their foolishness. It was then that Solomon turned the crown of gold into one of feathers. And it was a Hoopoe that brought reports of the Queen of Sheba to Solomon.

The Hoopoe is also found in Classical mythology. In Ovid's *Metamorphoses*, when Procne is transformed into a Swallow, and Philomela into a Nightingale (see p. 16), the violent King Tereus is turned into a Hoopoe with a crest representing his crown and a long sharp beak representing an offensive weapon.

According to the second-century AD text the *Physiologus*, Hoopoes would care for their elderly parents when they became infirm. As in ancient Egypt, Hoopoes were used medicinally, with Pliny the Elder stating that 'for pains in the side, the heart of a hoopoe is highly esteemed'.

In the twelfth-century mystical Persian poem by Farid ud-Din Attar, *The Conference of the Birds*, it is the Hoopoe that led a deputation of birds to seek knowledge from the legendary and mysterious bird, the Simorgh, and to make him their king. In this allegorical poem the party of birds cross seven valleys representing the stages that a man must pass to understand God: Yearning, Love, Gnosis, Detachment, Unity of God, Bewilderment, Selflessness and Oblivion in God. When the 30 birds reach the home of the Simorgh, all they see is themselves reflected in a lake. In the party, each bird represents a human fault, but the Hoopoe is wise and exemplifies qualities of leadership.

French folklore also gives the Hoopoe a reputation for cleverness. The Hoopoe and the woodpecker were friends and they journeyed across the sea together. But the woodpecker was tired, and the Hoopoe had to keep calling to keep his friend awake. When they reached land the woodpecker excavated a nest hole for the Hoopoe in thanks, and that explains why Hoopoes often nest in old woodpecker holes to this day, though they also nest in crevices in buildings or trees.

One less attractive aspect of Hoopoes is that they defend their nests with a foul-smelling liquid, which both adults and nestlings can squirt at intruders from their preen gland. They also smear this liquid, which smells like rotting meat, over their plumage, perhaps to act as an anti-parasite lotion as well as a further deterrent to predators. It is probably knowledge of this habit that led to the Hoopoe being included in Leviticus in the list of birds (as well as certain owls, among others) classified as 'detestable' and not to be eaten.

Despite it not being kosher, in 2008, Israel voted for the Hoopoe as its national bird. President Shimon Peres announced the result and lamented the fact that many once-common birds had become rare. He proclaimed the need for 'green scenery, fresh air and the beautiful, multicoloured birds' that flocked to the country.

For a foul-smelling bird, the Hoopoe enjoys a generally good reputation. It seems its boldness and jaunty walk, recognizable call and showy crest have endeared it to us wherever we live in association with it.

OPPOSITE
The Hoopoe speaks to the Peacock in *The Conference of the Birds*, a twelfth-century mystical Persian poem by Farid ud-Din Attar, in which the Hoopoe takes a leadership role, guiding the other bird species.

OVERLEAF
A Hoopoe nest is foul-smelling because the birds smear liquid from their preen gland over their plumage, and can also project it at any intruders. The nest illustrated here by late nineteenth-century Italian artist Eugenio Bettoni looks cleaner than most.

یار شد با من بیکجا مار زشت — تا بیفتادم بخواری از بهشت

چون بدل کردند خلوت جای من — تخته بند پای من شد پای من

عزم آن دارم کزین تاریک جای — رهبری باشد بخلدم رهنمای

من نه آن مرغم که در سلطان رسم — بس بود آنم که در دربان رسم

کی بود سیمرغ را پروای من — بس بود فردوس اعلی جای من

من ندارم در جهان کاری دگر — تا بهشتم ره دهد باری دگر

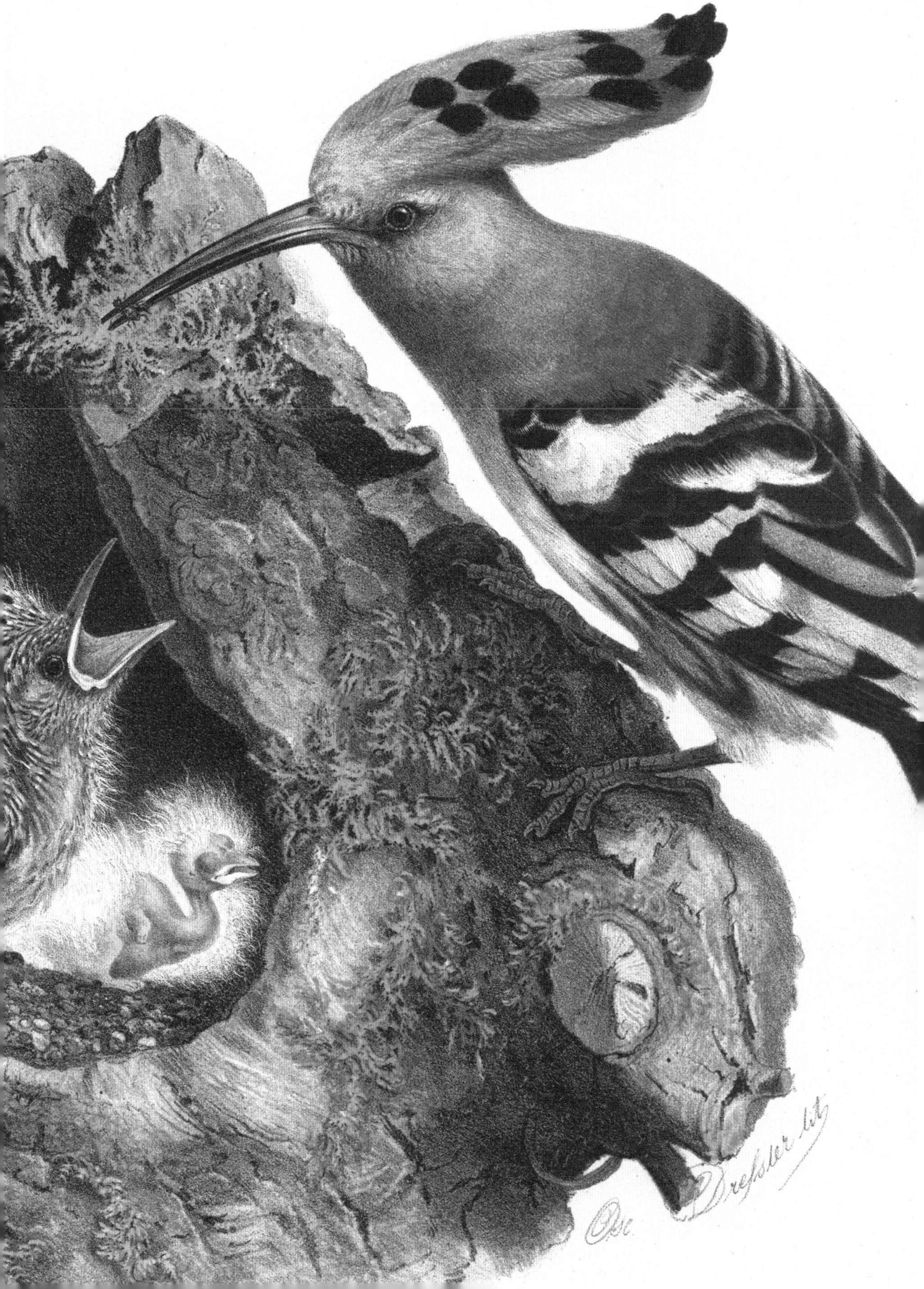

Further Reading

There has never been more information available on birds, their behaviour, ecology and place in our lives. Books about birds are published every week and the internet will take you to information about birds, where they occur, how to identify them and what makes them tick.

This section is not a complete reference list; it is an eclectic selection of internet sources and books that may be of interest to anyone wanting to find out more about birds and how remarkable they are. No scientific papers are listed here as they are less accessible for the average reader, both because they are often available only through subscription or in academic libraries and also because they are written for a professional and technical audience rather than the general reader. All those listed here are in the English language, and so there is an emphasis on UK and North American sources.

General

BOOKS

Birkhead, Tim R. *The Wisdom of Birds: An Illustrated History of Ornithology* (Bloomsbury, London and New York, 2008)

Birkhead, Tim R. *Bird Sense: What It's Like To Be a Bird* (Bloomsbury, London; Walker & Company, New York, 2012)

Birkhead, Tim R., Wimpenny, Jo and Montgomerie, Bob. *Ten Thousand Birds: Ornithology Since Darwin* (Princeton University Press, Princeton, 2015)

Campbell, Bruce, and Lack, Elizabeth. *A Dictionary of Birds* (T. & A. D. Poyser, Calton, 1985)

Carson, Rachel. *Silent Spring* (Houghton Mifflin, New York, 1962)

Cocker, Mark. *Birders: Tales of a Tribe* (Jonathan Cape, London; Atlantic Monthly Press, New York, 2001)

Cocker, Mark, and Mabey, Richard. *Birds Britannica.* (Chatto & Windus, London, 2005)

Cocker, Mark, and Tipling, David. *Birds and People* (Random House, London, 2014)

Del Hoyo, Josep. *The Handbook of Birds of the World* (17 volumes) (Lynx, Barcelona, 1992)

Elder, Charlie. *While Flocks Last* (Random House, London, 2009)

Elphick, Jonathan. *The World of Birds* (The Natural History Museum, London, 2014)

Grey, Edward. *The Charm of Birds* (Hodder & Stoughton, London; Frederick A. Stokes, New York, 1927)

Hauber, Mark E. and Bates, John. *The Book of Eggs: A Lifesize Guide to the Eggs of Six Hundred of the World's Bird Species* (Chicago University Press, Chicago, 2014)

Hume, Rob. *RSPB Complete Birds of Britain and Europe* (Dorling Kindersley, London, 2013)

The Illustrated Encyclopaedia of Birds (Dorling Kindersley, London, 2011)

Lack, David. *The Life of the Robin* (Penguin, London, 1953)

Lack, David. *Population Studies of Birds* (Oxford University Press, Oxford, 1966)

Lever, Christopher. *Naturalised Birds of the World* (T. & A. D. Poyser, London, 2005)

McCarthy, Michael. *Say Goodbye to the Cuckoo* (John Murray, London; Ivan R. Dee, Chicago, 2010)

Martin, Deborah L. *The Secrets of Backyard Bird-Feeding Success: Hundreds of Surefire Tips for Attracting and Feeding your Favorite Birds* (Rodale, New York, 2011)

Newton, Ian. *The Speciation and Biogeography of Birds* (Academic Press, Amsterdam and London, 2003)

Newton, Ian. *Bird Populations* (Collins, London, 2013)

Parr, Kevin. *The Twitch* (Unbound, London, 2014)

Scott, Peter. *The Eye of the Wind: An Autobiography* (Hodder, London; Houghton Mifflin, Boston, 1967)

Sibley, David. *The Sibley Guide to Birds* (Audubon Nature Guides Series, New York, 2008)

Snow, David W. and Perrins, Christopher M. *The Birds of the Western Palearctic (concise edition)* (Oxford University Press, Oxford and New York, 1997)

Soper, Tony. *Tony Soper's Bird Table Book: The Complete Guide to Attracting Birds and Other Wildlife to Your Garden* (David & Charles, Newton Abbot, 2006)

Svensson, Lars and Mullarney, Killian. *Collins Bird Guide* (HarperCollins, London, 2010)

Toft, Ron. *National Birds of the World* (Bloomsbury, London, 2013)

Wallace, Ian. *Beguiled by Birds* (Christopher Helm, London, 2004)

Westwood, Brett and Moss, Stephen. 2014. *Tweet of the Day: A Year of Britain's Birds from the Acclaimed Radio Series* (Saltyard, London, 2014)

WEBSITES

Here is a selection of general websites, many hosted and written by organizations that are worthy of your support in their work to conserve birds and educate the public about them.

10,000 birds
www.10000birds.com

ABA (American Birding Association)
www.aba.org

The American Ornithologists' Union
www.aou.org

Birdguides
www.birdguides.com

Birdlife International
www.birdlife.org

The British Ornithologists' Union
www.bou.org.uk

BTO
(British Trust for Ornithology)
www.bto.org

Cornell Laboratory
of Ornithology
www.birds.cornell.edu

National Audubon Society
www.audubon.org

RSPB
(Royal Society for
the Protection of Birds)
www.rspb.org.uk

WWT
(Wildfowl and Wetland Trust)
www.wwt.org.uk

MAGAZINES

Audubon
www.audubon.org

BBC Wildlife
www.discoverwildlife.com

Birdwatch
www.birdwatch.co.uk

Bird Watching
www.birdwatching.co.uk

British Birds
www.britishbirds.co.uk

British Wildlife
www.britishwildlife.com

North American Birds
www.aba.org/nab/

Songbirds

Beletsky, Les. *Bird Songs: 250 North American Birds in Song* (Chronicle, San Francisco, 2006)

Donald, Paul F. *The Skylark* (T. & A. D. Poyser, London, 2004)

Doughty, Robin W. *The Mockingbird* (University of Texas, Austin, 1995)

Elphick, Jonathan and Svennson, Lars. *Birdsong: 150 British and Irish birds and their songs* (Quadrille, London, 2012)

Feare, Chris. *The Starling* (Oxford University Press, Oxford and New York, 1984)

Gosler, Andy. *The Great Tit* (Hamlyn, London, 1993)

Kroodsma, Donald. *Singing Life of Birds: The Art and Science of Listening to Birdsong* (Houghton Mifflin, Boston, 2005)

Pepperberg, Irene Maxine. *The Alex Studies: Cognitive and Communicative Abilities of Grey Parrots* (Harvard University Press, Harvard, 2002)

Pepperberg, Irene Maxine. *Alex & Me* (Collins, New York, 2008)

Perrins, Christopher M. *British Tits* (Collins, London, 1979)

Rothenberg, David. *Why Birds Sing: A Journey into the Mystery of Birdsong* (Basic Books, New York, 2005)

WEBSITE

Xeno-canto has a wide range of bird songs and calls recorded from around the world www.xeno-canto.org

Birds of Prey

Baker, John A. *The Peregrine* (Collins, London; Harper & Row, New York, 1967)

Brown, Leslie. *British Birds of Prey* (Collins, London, 1976)

Brown, Philip. *Birds of Prey* (Andre Deutsch, London, 1964)

Dennis, Roy H. *A Life of Ospreys* (Whittles, Dunbeath, 2008)

Ferguson-Lees, James and Christie, David. *Raptors of the World* (Helm, London; Houghton Mifflin, Boston, 2001)

Gessner, David. *Return of the Osprey: A Season of Flight and Wonder* (Ballantine Books, London; Algonquin Books, Chapel Hill, 2002)

Glasier, Phillip. *Falconry and Hawking* (Batsford, London; Overlook Press, Woodstock, NY, 1998)

MacDonald, Helen. *H is for Hawk* (Jonathan Cape, London; Grove Press, New York, 2014)

Newton, Ian. *Population Ecology of Raptors* (T. & A. D. Poyser, Berkhamsted; Buteo Books, Vermillion, SD, 1979)

Newton, Ian. *The Sparrowhawk* (T. & A. D. Poyser, Calton, 1986)

Taylor, Marianne. 2010. *RSPB British Birds of Prey* (Bloomsbury, London, 2010)

Tingay, Ruth E. et al. (eds). *The Eagle Watchers: Observing and Conserving Raptors Around the World* (Comstock Publishing, Ithaca, NY, 2010)

Village, Andrew. *The Kestrel* (T. & A. D. Poyser, London, 2010)

Walter, Hartmut. *Eleonora's Falcon: Adaptations to Prey and Habitat in a Social Raptor* (University of Chicago Press, Chicago, 1979)

WEBSITES

The Hawk and Owl Trust www.hawkandowl.org

The International Centre of Birds of Prey www.icbp.org

The Peregrine Fund www.peregrinefund.org

Feathered Travellers

Davies, Nick. *Cuckoo: Cheating by Nature* (Bloomsbury, London, 2015)

Drewitt, Ed. *Urban Peregrines* (Pelagic, Exeter, 2014)

Dunphy, Madeleine. *The Peregrine's Journey: A Story of Migration* (Millbrook, Minneapolis, 2008)

Elphick, Jonathan (ed.). *Atlas of Bird Migration: Tracing the Great Journeys of the World's Birds* (The Natural History Museum, London; Firefly, Buffalo, NY, 2007)

Heinrich, Bernd. *The Homing Instinct: Meaning and Mystery in Animal Migration* (Houghton Mifflin, New York, 2014)

Lack, Andew J. and Overall, Roy. *The Museum Swifts: The Story of the Swifts in the Tower of the Oxford University Museum of Natural History* (Oxford University Museum of Natural History, Oxford, 2002)

Newton, Ian. *Bird Migration* (Collins, London, 2010)

Tilford, Tony. *The Complete Book of Hummingbirds* (Thunder Bay Press, San Diego, 2015)

The Love Life of Birds

Attenborough, David and Fuller, Errol. *Drawn from Paradise: The Discovery, Art and Natural History of the Birds of Paradise* (HarperCollins, London, 2012)

Birkhead, Mike and Perrins, Christopher M. *The Mute Swan* (Croom Helm, London, 1986)

Davies, Nicholas B. *Dunnock Behaviour and Social Evolution* (Oxford University Press, Oxford and New York, 1992)

Davies, Nicholas B. *Cuckoos, Cowbirds and other Cheats* (T. & A. D. Poyser, London; Academic Press, San Diego, 2000)

Davies, Nick. *Cuckoo: Cheating by Nature*. (Bloomsbury, London, 2015)

Fry, C. Hilary. *The Bee-eaters* (T. & A. D. Poyser, London, 1991)

Laman, Tim and Scholes, Edwin. *Birds of Paradise: Revealing the World's Most Extraordinary Birds* (National Geographic, Washington DC, 2012)

Ortega, Catherine. *Cowbirds and Other Brood Parasites* (University of Arizona Press, Tucson, 1998)

Parry, James. *The Mating Lives of Birds* (New Holland, London; MIT Press, Cambridge, MA, 2012)

Van Rhijn, Johan G. *The Ruff* (T. & A. D. Poyser, London; Academic Press, San Diego, 1991)

Avian Cities

Bruggers, Richard L. and Elliot, Clive C. H. (eds). *Quelea quelea: Africa's Bird Pest* (Oxford University Press, Oxford and New York, 1989)

Cherry-Garrard, Apsley. *The Worst Journey in the World* (Constable, London, 1922)

Davies, Lloyd S. *The Penguins* (T. & A. D. Poyser, London, 2003)

de Roy, Tui, Jones, Mark, and Cornthwaite, Julie. *Penguins: Their World, Their Ways* (Christopher Helm, London, 2013)

Harris, Mike P. and Wanless, Sarah. *The Puffin* (T. & A. D. Poyser, London, 2011)

Kear, Janet and DuPlaix-Hall, Nicole. *Flamingos* (T. & A. D. Poyser, London, 1975)

Nelson, J. Bryan. *Pelicans, Cormorants and Their Relatives* (Oxford University Press, Oxford and New York, 2005)

Shrubb, Michael. *The Lapwing* (T. & A. D. Poyser, London, 2007)

Taylor, Kenny. *Puffins* (Colin Baxter, Grantown-on-Spey, 1999)

Useful to Us

Baldwin, Suzie. *Chickens: The Essential Guide to Choosing and Keeping Happy, Healthy Hens* (Kyle, London, 2012)

Birkhead, Tim. *The Red Canary* (Bloomsbury, London, 2014)

Blechman, Andrew D. *Pigeons: The Fascinating Saga of the World's Most Revered and Reviled bird* (Grove, New York, 2007)

Clements, John. *Long-Distance Pigeon Racing* (Crowood Press, Marlborough, 2007)

Delacour, Jean. *The Pheasants of the World* (World Pheasant Association, Hindhead, 1977; Scribner, New York, 1951)

Diamond, Tony W. and Filion, F. I. (eds). *The Value of Birds* (International Council for Bird Preservation, Cambridge, 1987)

Dickson, James G. (ed.). *The Wild Turkey: Biology & Management* (Stackpole Books, Harrisburg, PA, 1992)

Goodwin, Derek. *Birds of Man's World* (British Museum (Natural History), London; Cornell University Press, Ithaca, 1978)

Gray, Nigel. *Woodland Management for Pheasants and Wildlife* (David & Charles, Newton Abbot, 1986)

Haupt, Thomas. *Canaries* (Barron's, Hauppauge, 2010)

Jerolmark, Colin. *The Global Pigeon* (University of Chicago Press, Chicago, 2013)

Johnsgard, Paul A. and Wolf, Joseph. *The Pheasants of the World* (Smithsonian Institution Press, Washington, DC, 1999)

Kear, Janet. *Man and Wildfowl* (T. & A. D. Poyser, London, 1990)

Threatened & Extinct

Avery, Mark. *A Message from Martha* (Bloomsbury, London, 2014)

Donald, Paul F., Collar, Nigel J., Marsden, Stuart J. and Pain, Deborah J. *Facing extinction: The World's Rarest Birds and the Race to Save Them* (T. & A. D. Poyser, London, 2013)

Fuller, Errol. *The Lost Birds of Paradise* (Swan Hill Press, Wykey, 1995)

Fuller, Errol. *The Great Auk* (E. Fuller, Southborough; Abrams, New York, 1999)

Fuller, Errol. *Dodo: From Extinction to Icon* (Collins, London, 2002)

Fuller, Errol. *Extinct Birds* (Comstock Publishing, Ithaca, NY, 2001)

Fuller, Errol. *The Passenger Pigeon* (Princeton University Press, Princeton, 2014)

Greenberg, Josh. *A Feathered River Across the Sky* (Bloomsbury, New York, 2014)

Hume, Julian P. and Walters, Michael. *Extinct Birds* (T. & A. D. Poyser, London, 2012)

Juniper, Tony. *Spix's Macaw: The Race to Save the World's Rarest Bird* (Fourth Estate, London and New York, 2002)

Schorger, Arlie W. *The Passenger Pigeon* (University of Oklahoma Press, Norman, 1955)

Snyder, Noel F. R. and Snyder, Helen. *The California Condor: A Saga of Natural History and Conservation* (Academic Press, New York, 2000)

Steadman, Ralph and Levy, Ceri. *Extinct Boids* (Bloomsbury, London, 2012)

WEBSITE

Saving the spoon-billed sandpiper www.saving-spoon-billed-sandpiper.com

Revered & Adored

Heinrich, Bernd. *Mind of the Raven: Investigations and Adventures with Wolf-Birds* (Cliff Street Books, New York, 1999)

Parry-Jones, Jemima. *Eagles and Birds of Prey* (Dorling Kindersley, London, 1997)

Ratcliffe, Derek A. *The Raven* (T. & A. D. Poyser, London, 1997)

Rowland, Beryl. *Birds with Human Souls: A Guide to Bird Symbolism* (University of Tennessee Press, Knoxville, 1978)

Savage, Candace. *Crows: Encounters with the Wise Guys of the Avian World* (Greystone Books, Vancouver, 2005)

Sewell, Matt. *Owls: Our Most Enchanting Birds* (Ebury, London, 2014)

Tingay, Ruth E. et al. (eds). *The Eagle Watchers: Observing and Conserving Raptors around the world* (Comstock Publishing, Ithaca, NY, 2010)

Toms, Mike. *Owls* (Collins, London, 2014)

Sources of Illustrations

All images are from the collections of the British Library unless otherwise stated.

Eggs are from S. L. Mosley, *A History of British birds, their nests, and eggs. With hand-coloured figures of all the species and varieties*, 1884. **2.** Cornelius Nozeman, *Nederlandsche Vogelen*, 1770–1829. **4.** Skylark: Cornelius Nozeman, *Nederlandsche Vogelen*, 1770–1829. **5.** Nightingale, Great Tit and Skylark: John Selby Prideaux, *Atlas: Illustrations of British Ornithology*, 1821–34; Swift: Cornelius Nozeman, *Nederlandsche Vogelen*, 1770–1829; Eleonora's Falcon: Charles Robert Bree, *A History of the Birds of Europe, not observed in the British Isles*, 1859; Secretary Bird: William Hayes, *Portraits of Rare and Curious Birds, with their descriptions, from the Menagerie of Osterly Park*, 1794–99; Goldcrest: Francis Orpen Morris, *A History of British Birds*, 1870; Hummingbird: William Swainson, *The ornithological drawings of W.S. (Series 1.) The Birds of Brazil*, 1834–35. **6.** Puffin: Francis Orpen Morris, *A History of British Birds*, 1870; Pelican: John James Audubon, *The Birds of America*, 1827–38; Pigeon: Robert Fulton, *The Illustrated Book of Pigeons, illustrated with plates from paintings by J. W. Ludlow*, 1874–76; European Bee-eater and Bullfinch: John Selby Prideaux, *Atlas: Illustrations of British Ornithology*, 1821–34; Turkey: Eleazar Albin, *A Natural History of Birds*, 1738; Pheasant: Beverley Robinson Morris, *British Game Birds and Wildfowl*, 1895. **7.** Moa: Walter Rothschild, *Extinct Birds*, 1907; Great Auk: Francis Orpen Morris, *A History of British Birds*, 1870; Jungle Fowl: Heinrich Rudolf Schinz, *Naturgeschichte und Abbildungen der Vögel: gezeichnet und lithographirt von K. J. Brodtmann*, 1836; Chatham Island Black Robin, painting by Janet E. Marshall, www.janetmarshall.co.nz; Dodo: Walter Rothschild, *Extinct Birds*, 1907; Tawny Owl and Hoopoe: John Selby Prideaux, *Atlas: Illustrations of British Ornithology*, 1821–36. **8.** John James Audubon, *The Birds of America*, 1827–38. **9.** John Selby Prideaux, *Atlas: Illustrations of British Ornithology*, 1821–35. **10.** John Gould, *Birds of Australia*, 1848–69; William Swainson, *The ornithological drawings of W.S. (Series 1.) The Birds of Brazil*, 1834–35. **11.** John Selby Prideaux, *Atlas: Illustrations of British Ornithology*, 1821–35; Archibald Thorburn, *British Birds*, 1915–18. **12.** John Selby Prideaux, *Atlas: Illustrations of British Ornithology*, 1821–35; *Chorui Hiako Shiu'*, Japan, *c.* 1860–70, Or.5450, f.3v. **13.** George Robert Gray, *The Genera of Birds. Illustrated with ... plates by D. W. Mitchell*, 1844–49. **14.** Wood-engraving by Thomas Bewick, *c.* 1791–97, British Museum, London. **16.** Chromolithograph by J. G. Keulemans, Thomas Littleton Powys Lilford, Baron, *Coloured Figures of the Birds of the British Islands*, 1895–97. **17.** Eugenio Bettoni, *Storia naturale degli uccelli che nidificano in Lombardia, ad illustrazione della raccolta ornitologica dei fratelli Ercole ed Ernesto Turati ... Con tavole litografate e colorate, prese dal vero da O. Dressler*, 1865–71. **18.** J. A. Naumann, *Naturgeschichte der Vögel Mittel-Europas. ... Herausgegeben von Dr. C. R. Hennicke, etc.*, 1905. **19.** J. G. Keulemans, *Onze Vogels in huis entuin, beschreven en afgebeeld door J. G. K.*, 1869–76. **21.** Eugenio Bettoni, *Storia naturale degli uccelli che nidificano in Lombardia, ad illustrazione della raccolta ornitologica dei fratelli Ercole ed Ernesto Turati ... Con tavole litografate e colorate, prese dal vero da O. Dressler*, 1865–71. **22.** J. A. Naumann, *Naturgeschichte der Vögel Deutschlands*, 1822–60. **23.** Illustration by Louis Agassiz Fuertes, Charles Benedict Davenport, *Introduction to zoology; a guide to the study of animals, for the use of secondary schools*, 1911. **24.** John James Audubon, *The Birds of America*, 1827–38. **25.** Thornton Waldo Burgess, *The Burgess Bird Book for Children ... With illustrations in color by Louis Agassiz Fuertes*, 1919. **26.** John Selby Prideaux, *Atlas: Illustrations of British Ornithology*, 1821–34. **27.** Watercolour by Chobi, Heyne Collection, NHD1/58. **27.** Watercolour by Chobi, Heyne Collection, NHD1/33. **28.** Francis Orpen Morris, *A History of British Birds*, 1870. **29.** Eugenio Bettoni, *Storia naturale degli uccelli che nidificano in Lombardia, ad illustrazione della raccolta ornitologica dei fratelli Ercole ed Ernesto Turati ... Con tavole litografate e colorate, prese dal vero da O. Dressler*, 1865–71. **30.** J. A. Naumann, *Naturgeschichte der Vögel Deutschlands*, 1822–60. **31.** J. G. Keulemans, *A Natural history of cage birds*, 1871. **33.** Saverio Manetti, *Storia naturale degli uccelli trattata con metodo ... Ornithologia methodice digesta, etc.*, 1767–76. **34.** John Selby Prideaux, *Atlas: Illustrations of British Ornithology*, 1821–34. **36.** Eugenio Bettoni, *Storia naturale degli uccelli che nidificano in Lombardia, ad illustrazione della raccolta ornitologica dei fratelli Ercole ed Ernesto Turati ... Con tavole litografate e colorate, prese dal vero da O. Dressler*, 1865–71. **37.** J. G. Keulemans, *Onze Vogels in huis entuin, beschreven en afgebeeld door J. G. K.*, 1869–76. **39.** Wellesley Collection, NHD29/96. **40.** *Transactions of the Linnean Society*, Vol.VI, 1802. **42.** John James Audubon, *The Birds of America*, 1827–38. MEPL/Natural History Museum. **43.** Mark Catesby, *The natural history of Carolina, Florida and the Bahama Islands...*, 1731–43. **44.** British Museum, London. **46.** The Art Archive/Alan Brook/NGS Image Collection. **47.** John Selby Prideaux, *Atlas: Illustrations of British Ornithology*, 1821–34. **48.** Archibald Thorburn, *British Birds*, 1915–18. **50.** Lithograph by Joseph Wolf, *c.* 1860, *Transactions of the Zoological Society of London*, Vol VI. 1866, MEPL/Natural History Museum. **51.** The Art Archive/Allan Brook/NGS Image Collection. **52.** Charles Robert Bree, *A History of the Birds of Europe, not observed in the British Isles*, 1859. **53.** J. A. Naumann, *Naturgeschichte der Vögel Mittel-Europas. ... Herausgegeben von Dr. C. R. Hennicke, etc.*, 1905. **54.** Lithograph by Joseph Wolf, *The Ibis*, ser. 2, vol. 5., 1869. **55.** Jemima Blackburn, *Birds drawn from Nature by Mrs. Hugh Blackburn*, 1868. **56.** Eugenio Bettoni, *Storia naturale degli uccelli che nidificano in Lombardia, ad illustrazione della raccolta ornitologica dei fratelli Ercole ed Ernesto Turati ... Con tavole litografate e colorate, prese dal vero da O. Dressler*, 1865–71. **57.** *Birds of Prey. [Twelve cards with descriptive letterpress.]*, 1870. **59.** Harry Swann, *Monograph of Birds of Prey*, 1924. **60.** William Swainson, *Selection of the Birds of Brazil and Mexico*, 1841. **61.** *Birds of Prey. [Twelve cards with descriptive letterpress.]*, 1870. **62.** George Shaw, *The Naturalists' Miscellany, or coloured figures of natural objects drawn and described ... from nature*, 1751–1813. **63.** Album of natural history drawings belonging to Joseph Banks, Add. Ms.11807, f.15. **64-65.** *Album of Abyssinian Birds and Mammals. From paintings by Louis Agassiz Fuertes*, 1930. **66.** Patrick Guenette/123RF. **68.** Cornelius Nozeman, *Nederlandsche Vogelen*, 1770–1829. **70.** Francis Orpen Morris, *A History of British Birds*, 1870. **71** The Art Archive/Allan Brooks/NGS Image Collection. **72.** Francis Orpen Morris, *A History of British Birds*, 1870. **73, 74.** J. A. Naumann, *Naturgeschichte der Vögel Deutschlands*, 1822–60. **75.** Jemima Blackburn, *Birds drawn from Nature by Mrs. Hugh Blackburn*, 1868. **77.** John James Audubon, *The Birds of America*, 1827–38. **78.** Heinrich Rudolf Schinz, *Naturgeschichte und Abbildungen der Vögel: gezeichnet und lithographirt von K. J. Brodtmann*, 1836. **79.** John Gould, *An introduction to the Trochilidae, or family of humming-bird*, 1861. **80.** William Swainson, *The ornithological drawings of W.S. (Series 1.) The Birds of Brazil*, 1834–35. **81** Robert John Thornton, *Temple of Flora*, 1807. **82.** Raja of Tangore Collection, NHD2/1011. **83.** Harry Swann, *Monograph of Birds of Prey*, 1924. **84.** Louis Agassiz Fuertes, *Birds of America*, 1940. **85.** Hermann Schlegel and A. H. Verster van Wulverhorst, *Traité de Fauconnerie*, 1844–53. **86.** Patrick Guenette/123RF. **89.** Archibald Thorburn, *British Birds*, 1915–18. **90.** The Art Archive/Victoria & Albert Museum. **91.** William MacGillivray, *Watercolour drawings of British Animals*, 1831–41, MEPL/Natural History Museum. **93.** Gregory Macalister Mathews, *The Birds of Norfolk & Lord Howe Islands and the Australasian South Polar Quadrant. With additions to "The Birds of Australia," etc.*, 1928–36. **94.** Watercolour by Frank L. Beebe, 1957, Royal BC Museum, Victoria, Canada. **95.** M. Henri A. de Conty, *Types du règne animal. Buffon en estampes. Planches & illustrations par M. É. Traviès*, 1862–64. **96, 97.** François Le

Vaillant, *Histoire naturelle des oiseaux de Paradis et des rolliers*, Vol.1, 1806. **98, 99.** John Gould, *The Birds of New Guinea and the adjacent Papuan Islands, including any new species that may be discovered in Australia*, 1875. **100.** Francis Willughby, *The Ornithology of F. W. ...*, 1678. **101.** *Chorui Hiako Shiu'*, Japan, *c.* 1860–70, Or.5450. **103.** Eugenio Bettoni, *Storia naturale degli uccelli che nidificano in Lombardia, ad illustrazione della raccolta ornitologica dei fratelli Ercole ed Ernesto Turati ... Con tavole litografate e colorate, prese dal vero da O. Dressler*, 1865–71. **104.** Pencil & watercolour, painted from life at Wells-next-the-Sea, Norfolk, 15 March 2007, © James McCallum. **105.** J. A. Naumann, *Naturgeschichte der Vögel Mittel-Europas. ... Herausgegeben von Dr. C. R. Hennicke, etc.*, 1905. **106.** Eugenio Bettoni, *Storia naturale degli uccelli che nidificano in Lombardia, ad illustrazione della raccolta ornitologica dei fratelli Ercole ed Ernesto Turati ... Con tavole litografate e colorate, prese dal vero da O. Dressler*, 1865–71. **107.** John Selby Prideaux, *Atlas: Illustrations of British Ornithology*, 1821–34. **108.** John James Audubon, *The Birds of America*, 1827–38. **109** The Art Archive/Allan Brook/NGS Image Collection. **111** John Selby Prideaux, *Atlas: Illustrations of British Ornithology*, 1821–34. **112.** Cornelius Nozeman, *Nederlandsche Vogelen*, 1770–1829. **113** Thomas Pennant, *British Zoology*, 1766. **114-15.** J. A. Naumann, *Naturgeschichte der Vögel Mittel-Europas. ... Herausgegeben von Dr. C. R. Hennicke, etc.*, 1905. **116.** Ulyssis Aldrovandi, *Ornithologiae ... libri XII (XX), etc.*, 1646. **118.** J. A. Naumann, *Naturgeschichte der Vögel Mittel-Europas. ... Herausgegeben von Dr. C. R. Hennicke, etc.*, 1905. **120.** Watercolour by John White, 1585–93, British Museum, London. **121.** Eleazar Albin, *A Natural History of Birds. Illustrated with ... copper plates ... colour'd, etc. [With observations by W. Derham.]*, 1738. **122.** Ulyssis Aldrovandi, *Ornithologiae ... libri XII (XX), etc.*, 1646. **123.** John James Audubon, *The Birds of America*, 1827–38. **124.** The Art Archive/Walter A. Weber/NGS Image Collection. **125.** Watercolour by Guru Dayal, Gibbons & Buchanan Collection, NHD 2/278. **126.** S. L. Mosley, *A History of British birds, their nests, and eggs. With hand-coloured figures of all the species and varieties*, 1884. **127.** J. G. Keulemans, *Onze Vogels in huis entuin, beschreven en afgebeeld door J. G. K.*, 1869–76. **128.** J. A. Naumann, *Naturgeschichte der Vögel Mittel-Europas. ... Herausgegeben von Dr. C. R. Hennicke, etc.*, 1905. **129.** Alexei Kondratyevich Savrasov, *The Rooks are Back Again*, oil on canvas, 1871, Tretyakov Gallery, Moscow. **130.** Thomas Pennant, *British Zoology*, 1766. **131.** William Yarrell, *History of British Birds*, 1871–85. **132.** © Estate of Charles Tunnicliffe. **133.** Chromolithograph by Archibald Thornburn, Thomas Littleton Powys Lilford, *Coloured figures of the Birds of the British Islands*, 1895–97. **134.** Scott Polar Research Institute, University of Cambridge, UK/Bridgeman Images. **135.** Album of natural history drawings belonging to Joseph Banks, Add. Ms.11807, f.26. **136.** John Richardson and John Edward Gray, *The zoology of the voyage of the H.M.S. Erebus & Terror*, 1844. **137-38.** Scott Polar Research Institute, University of Cambridge, UK/Bridgeman Images. **140, 142.** Ulyssis Aldrovandi, *Ornithologiae ... libri XII (XX), etc.*, 1646. **143.** Drawing by an Indian artist, Punjab, *c.* 1840, NHD41/54. **144.** Welleseley Collection, NHD29/69. **145.** Woodblock print by Hiroshige, *c.* 1832, British Museum, London. **146.** Painting by Mansur, *c.* 1612, The Art Archive/Victoria & Albert Museum. **148.** Eleazar Albin, *A Natural History of Birds*, 1738. **149.** J. A. Naumann, *Naturgeschichte der Vögel Mittel-Europas. ... Herausgegeben von Dr. C. R. Hennicke, etc.*, 1905. **150.** John Gould, *Birds of Asia*, 1850–83. **151.** Archibald Thorburn, *British Birds*, 1915–18. **152.** *Chorui Hiako Shiu'*, Japan, *c.* 1860–70, Or.5450. **153.** Francis Orpen, Morris, *A History of British Birds*, 1870. **154.** Suzuki Harunobu, *Cormorant Fishing by Night*, woodblock print, 1724–70. British Museum, London. **155.** Hanging scroll by Katsushika Hokusai, 1847, British Museum, London. **156.** Gino Rossi, *The Man with the Canary*, oil on canvas, 1913, The Art Archive/Private Collection Milan/Mondadori Portfolio/Electa. **157.** W. A. Blakston, W. Swaysland and A. F. Wiener, *The Illustrated Book of Canaries, and Cage-Birds, British and Foreign*, 1877–80. **158.** C. J. Temminick, *Nouveau recueil de planches coloriées d'oiseaux*, 1838. **159.** John Gould, *Birds of Asia*, 1850–83. **160.** Robert Fulton, *The Illustrated book of Pigeons, illustrated with plates from paintings by J. W. Ludlow*, 1874–76. **161.** After a drawing by Edward Lear, John Selby Prideaux, *The Natural History of Pigeons*, 1835. **162-63.** Eugenio Bettoni, *Storia naturale degli uccelli che nidificano in Lombardia, ad illustrazione della raccolta ornitologica dei fratelli Ercole ed Ernesto Turati ... Con tavole litografate e colorate, prese dal vero da O. Dressler*, 1865–71. **164.** Francis Willughby, *The Ornithology of F. W. ...*, 1678. **166.** Walter Lawry Buller, *Buller's birds of New Zealand*, 1873. **167.** Walter Rothschild, *Extinct Birds*, 1907. **168.** Cornelis Saftleven, *Dodo*, watercolour, 1638, Museum Boijmans van Beuningen, Rotterdam. **169.** George Shaw, *The Naturalists' Miscellany, or coloured figures of natural objects drawn and described ... from nature*, 1751–1813. **170.** Walter Rothschild, *Extinct Birds*, 1907. **171** Francis Orpen Morris, *A History of British Birds*, 1870. **172.** Illustration by Louis Agassiz Fuertes, William Buttes Mershon, *The Passenger Pigeon*, 1907. **173** John James Audubon, *The Birds of America*, 1827–38. **174, 175.** Painting by Janet E. Marshall, www.janetmarshall.co.nz. **176.** John Gould, *A synopsis of the Birds of Australia and the adjacent islands*, 1837–38. **177.** Walter Lawry Buller, *Buller's birds of New Zealand*, 1873. **179.** John James Audubon, *The Birds of America*, 1827–38. **180.** Mark Catesby, *The natural history of Carolina, Florida and the Bahama Islands:...*, 1731–43. **183.** Edward Lear, *Illustrations of the family of psittacidae or parrots*, 1832. **184.** Thomas Dick Lauder, *The Miscellany of Natural History*, 1833–34. **185.** Charles de Souancé, *Iconographie des perroquets, non figurés dans les publications de Levaillant et de Bourjot Saint-Hilaire*, 1837–38. **187.** John Gould, *Birds of Asia*, 1850–83. **188.** John James Audubon, *The Birds of America*, 1827–38. **189.** George Robert Gray, *The Genera of Birds. Illustrated with ... plates by D. W. Mitchell*, 1844–49. **190.** George Shaw, *The Naturalists' Miscellany, or coloured figures of natural objects drawn and described ... from nature*, 1751–1813, MEPL/Natural History Museum. **191.** James Wilson, *Illustration of Zoology, being representations of new, rare or remarkable subjects of the animal kingdom*, 1831. **192.** George Robert Gray, *The Genera of Birds. Illustrated with ... plates by D. W. Mitchell*, 1844–49. **193.** Jacobi Christiani Schaeffer, *Elementa Ornithologica iconibus ... illustrate*, 1774. **194-95.** John James Audubon, *The Birds of America*, 1827–38. **196.** Francis Willughby, *The Ornithology of F. W. ...*, 1678. **198.** Woodblock print by Kawanabe Kyosai, British Museum, London. **199.** J. A. Naumann, *Naturgeschichte der Vögel Mittel-Europas. ... Herausgegeben von Dr. C. R. Hennicke, etc.*, 1905. **200.** Northwest coast shaman's rattle, British Museum, London. **201.** Watercolour by Haludar, Gibbons & Buchanan collection, NHD2/278; Lithograph by Joseph Wolf, *The Ibis*, ser. 2, vol. 5., 1869. **202.** Vivant Denon, *Momies d'Ibis*, pen and brown ink, brown and grey wash, *c.* 1802, British Museum, London. **203.** *The Ibis*, ser. 2, vol. 5, 1869. **204.** Carel Fabritius, *The Goldfinch*, oil on canvas, 1654, Mauritshaus, Netherlands, akg images/Andre Held. **205.** Eugenio Bettoni, *Storia naturale degli uccelli che nidificano in Lombardia, ad illustrazione della raccolta ornitologica dei fratelli Ercole ed Ernesto Turati ... Con tavole litografate e colorate, prese dal vero da O. Dressler*, 1865–71. **206.** Carlo Crivelli, *Madonna and Child*, Tempera and gold on wood, *c.* 1480, Metropolitan Museum of Art, New York. **207.** *The Sherborne Missal*, early 15th century, Add.Ms.74236, p. 396. **208.** Bolivian mask, British Museum, London. **209.** Harry Swann, *Monograph of Birds of Prey*, 1924. **210.** Edward Whymper, *Travels amongst the Great Andes of the Equator*, 1892. **211.** Engraving by Gustave Doré, Samuel Taylor Coleridge, *The Rime of the Ancient Mariner*, 1877. **212.** F. H. H. Guillemard, *Ocean Birds*, 1887. **213.** Walter Lawry Buller, *Buller's birds of New Zealand*, 2nd edition, 1888. **214.** John James Audubon, *The Birds of America*, 1827–38. **215** Eleazar Albin, *A Natural History of Birds*, 1738. **216.** Musée du Beaux Arts, Lyon. **217.** Cornelius Nozeman, *Nederlandsche Vogelen*, 1770–1829. **219.** *Album of Abyssinian Birds and Mammals. From paintings by Louis Agassiz Fuertes*, 1930. **220.** John Selby Prideaux, *Atlas: Illustrations of British Ornithology*, 1821–34. **221.** Mark Catesby, *The natural history of Carolina, Florida and the Bahama Islands...*, 1731–43. **222.** Woodblock print by Katsushika Hokusai, British Museum, London. **224.** James Wilson, *Illustration of Zoology, being representations of new, rare or remarkable subjects of the animal kingdom*, 1831; *Codex Mendoza, the Mexican Manuscript known as the Collection of Mendoza and preserved in the Bodleian Library, Oxford*, 1938. **225.** John Gould, *A Monograph of the Trogonidæ, or family of Trogons*, 1838. **226.** John Gould, *Birds of Great Britain*, 1862–73. **227.** Francis Orpen Morris, *A History of British Birds*, 1870. **229.** *Mantiq al-tayr (The Conference of the Birds) of 'Attar*, Add.Ms.7735, f. 30v. **230-31.** Eugenio Bettoni, *Storia naturale degli uccelli che nidificano in Lombardia, ad illustrazione della raccolta ornitologica dei fratelli Ercole ed Ernesto Turati ... Con tavole litografate e colorate, prese dal vero da O. Dressler*, 1865–71.

Index

Bold page numbers indicate main illustrated entries for bird species. *Italic* page numbers refer to illustrations.

O

P

Q

R

S

T

U

V

W

Y

Z

About the Author

Dr Mark Avery is a scientist and naturalist who writes about and comments on environmental issues. He worked for the RSPB for 25 years and was the RSPB's Conservation Director for nearly 13 years. His most recent books include *Fighting for Birds: 25 Years in Nature Conservation* (2012), *A Message from Martha: The Extinction of the Passenger Pigeon and Its Relevance Today* (2014), *Behind the Binoculars: Interviews with Acclaimed Birdwatchers* (with Keith Betton; 2015) and *Inglorious: Conflict in the Uplands* (2015). He also contributes to numerous journals and magazines.